General Practical Chemistry

General Practical Chemistry
(for B.Sc. and Pre-Medical courses)

Prof. (Dr.) Marei M. El-ajaily
Professor of Chemistry,
Chemistry Department, Faculty of Science,
Benghazi University, Benghazi, Libya

Prof. (Dr.) Abdussalam A. Maihub
Professor of Chemistry,
Chemistry Department, Faculty of Science,
Tripoli University, Tripoli, Libya

Dr. Ranjan K. Mohapatra
Former Head, Department of Basic Science &
Humanities,
Government College of Engineering,
Keonjhar, Odisha, India

CWP
Central West Publishing

NATIONAL LIBRARY OF AUSTRALIA

A catalogue record for this book is available from the National Library of Australia

ISBN (print): 978-1-925823-78-3

Contents

Preface

This book presents the general principles of practical chemistry for the students of B.Sc. and premedical courses. We believe that this book will be very useful for all students reading chemistry. The topics included in this textbook will help the students to understand the principles and experimental procedures in inorganic chemistry, organic chemistry and analytical chemistry. This book contains basic laboratory techniques, qualitative inorganic analysis, qualitative organic analysis, synthesis of some organic compounds, quantitative (volumetric and gravimetric) analysis of chemical compounds and synthesis of some coordination compounds. We hope this book will be very useful and a good offered book for our students and readers. Authors tried to present this practical book in simple and lucid manner.

The authors are grateful to the editorial team of Central West Publishing, Australia for their excellent work and constructive suggestions, which helped in successfully publishing this book. The authors welcome any suggestions and criticisms for improvement of the book.

Prof. (Dr.) Marei M. El-ajaily
Prof. (Dr.) Abdussalam A. Maihub
Dr. Ranjan K. Mohapatra

Chapter I

General Introduction

I.1 General rules for the laboratory

The organic compounds are dangerous to work with because of their inflammability and toxicity. Also, the organic compounds are volatile enough to cause fires, especially when naked Bunsen flames are around. However, there are simple rules which should be followed to ensure safety. A chemistry laboratory may be one of the most dangerous work areas known, especially when the laboratory is not treated with respect. In this section, we will discuss the safety procedures that must be used.

It is very important to be aware of people and facilities needed when a laboratory accident results that demands expert assistance. For this, the following phone numbers should be available to the laboratory instructor so that expert assistance can be quickly solicited: Ambulance, Fire department, Institute physician, Institute health center, Police, Poison control center, etc.

I.2 General rules

Read the following instructions carefully:

- Keep your position neat and clean.
- Always use an apron in the laboratory.
- Do not drink in the laboratory.
- Do not taste any chemicals (even NaCl, which is called common salt).
- Do not ignite a Bunsen burner until you are sure that there is no inflammable vapour nearby.
- Make sure that you know where the nearest fire extinguisher is installed and how it works.
- If a small fire (e.g. in a beaker) starts, it may be extinguished by placing rags over it, to exclude air.
- Do not throw material which is wet with solvent into the waste bin.
- If you spill an organic compound on your skin, wipe it off immediately, and then wash with running tap water.
- Vapours of organic solvents are toxic. Do not evaporate solvents into the laboratory atmosphere, always use a rotary evaporator to

condense the vapour and collect the liquid for disposal in the solvent residue bottles.

N.B.: The following common chemicals are dangerous. Hence, these should be used in a fume cupboard with adequate ventilation.

I.3 Some common poisons and their effects

Name of the chemicals	Effects
Acids and alkalis	will burn and corrode the tissues in contact
Alcohol	acts as a strong depressant on the central nervous system; also affects a person's stomach
Cyanide	lower concentrations may make the victim unconscious, followed by death because of respiratory paralysis
Carbon monoxide	combines with oxygen-carrying system of blood
Hydrogen sulfide	flammable and poisonous gas with odor of rotten eggs; lower concentrations may cause irritation of mucous membranes, headache, nausea and lassitude
Lead	chronic cases may show weight loss, weakness and anemia
Mercury	dangerous because it is fairly volatile and readily absorbed by the intact skin, respiratory tract and gastrointestinal tract; causes vomiting, abdominal pain, kidney damage and death within 10 days
Methyl alcohol	may cause permanent damage and blindness on contact
Phenyl hydrazine	causes hemolytic reaction with erythrocytes
Pyrethrin	found in insecticides; may cause paralysis of general muscles and respiratory actions
Silver nitrate	contact with skin or mucous membranes may be caustic and irritating; Swallowing may cause severe gastroenteritis and eventually death
Benzene	extremely toxic as an acute or chronic poison; commonly used as a solvent, but due to its

	toxic nature, toluene may be used as a substitute
Carbon tetrachloride	acute exposure results in central nervous system depression followed by hepatic and renal damage; Large doses can cause death
Chloroform and dichloromethane	act as central nervous system depressant

One should check the list of poisons in order to decide whether vomiting should be induced or not. In most of these cases vomiting would result in passing the corrosive poison back through delicate bodily tissues.

I.4 Poisons for which vomiting should not be induced

If a person has swallowed the following chemicals, give milk or water as per age.

Ammonia	Calcium oxide (Lime)	Benzene
Sodium hydroxide	Sodium hypochlorite (Bleach)	Petroleum ether (Naphtha)
Carbolic acid	Paint thinners and removers	Phenols
Pine oil	Detergents	Sodium carbonate
Dry cleaning fluids	Strong acids	Gasoline
Strychnine	Kerosene	Sodium bicarbonate (washing soda)

I.5 Poisons for which vomiting should be induced

If a person has swallowed the following chemicals, vomiting should be induced.

Alcohol (rubbing, ethyl)	Alcohol (wood, methyl)
Borax	Formaldehyde antifreeze (ethylene glycol)
Camphor	Insect repellents

I.6 Handy first aid tips

Wounds
Objective: Protect the wound from infection and control bleeding.
First aid: Use sterile dressings and apply direct pressure on the wound until bleeding stops.

Shock
Objective: Keep lying down and comfortable.
Symptoms: Moist pale skin, shallow breathing, lack luster eyes, in addition to weakness.
First aid:
1. Keep lying down and elevate feet if no head or chest injury.
2. Cover with blanket (do not cause sweating).
3. Give water to allay thirst.

Artificial respiration
Objective: Open airways, keep airways open, alternating increase and decrease in size of chest.
Symptoms: Not breathing from electric shock, drowning, gas poisoning.
First aid:
1. Pull jaw forward and tilt head back.
2. Seal victim's nose and mouth with your mouth and breathe directly into victim's mouth.
3. Remove your mouth and turn head while victim exhales.
4. Repeat 15-20 times per minute.

Poisons
Objective: Dilute the poison and induce vomiting, except as advised.
Symptoms: Burns around mouth, empty bottle, etc.
First aid:
1. Dilute with H_2O or milk.
2. Induce vomiting by strong baking soda solution or finger.
3. **Universal antidote**: part strong tea, part of milk of magnesia and 2 parts burnt toast.
4. Do not cause vomiting if victim has swallowed strong acid, kerosene or strychnine. Check all labels for antidote.

Fractures
Objective: Keep broken bone ends and adjacent joints fixed.
Symptoms: Pain, swelling, deformity, etc.

First aid: Use stiff material pillow or blanket and splint, whatever is available.

Burns
Objective: Relieve pain and prevent infection.
Symptoms: 1st degree = redness; 2nd degree = blisters; 3rd degree = deep tissue damage
First aid:
1. Cover with thick layer of sterile dry dressing.
2. Wash with water for chemical burns.

Transportation
Objective: If moving the victim is necessary, do not bend, twist or shake the victim.   .
First aid:
1. Drag victim on coat, blanket or rug.
2. Use chair, stretcher or carry manually.
3. Do not induce any further injury to the victim.

Fainting: Have person lie flat if possible or have person lower head between knees and breathe deeply. Use ammonia ampoule, if available.

Heart attack: If person has medication, assist in giving the medicine; keep lying down; allow for comfortable breathing; call the physician.

1.7 Safety instructions

Eye protection

As the eyes are particularly susceptible to permanent damage by corrosive chemicals as well as flying objects, *safety goggles* must be worn at all times in the laboratory. *Prescription glasses* are not recommended since they do not provide a proper side protection. *No sunglasses* are allowed in the laboratory. *Contact lenses* are a potential hazard because the chemical vapours dissolve in the liquid covering the eye and concentrate behind the lenses. If you have to wear contact lenses, consult with your instructor. If possible, try to wear a prescription glasses under your safety goggles. In case of any accident involving a chemical splash near your eyes, immediately wash your eyes with copious amount of water and inform your instructor. Especially, when heating a test tube, do not point its mouth to anyone.

Cuts and burns

Use glassware with care. Glassware is breakable and may cause cuts. If a piece of glass is heated it may get heated very quickly. As hot glass is not disbursable from the cold one, handle these with a tong. Do not use any cracked or broken glass equipment. It may ruin an experiment and worse, it may cause serious injury. Place damaged glass in a waste glass container. Do not throw broken glass in the waste paper container or regular waste container.

Poisonous chemicals

Never taste any chemicals in the laboratory unless specifically directed to do so. If you are asked to taste the odour of a substance, do it by *wafting* a bit of the vapour toward your nose. Do not stick your nose in and inhale vapour directly from the test tube. Always wash your hands before leaving the laboratory. Eating and drinking any type of food are prohibited in the laboratory at all times. Smoking in the laboratory premises is not allowed.

Clothing and footwear

Everyone must wear **a lab coat** inside the lab. Long hair should be securely tied back to avoid the risk of fire. If large amounts of chemicals are spilled on your body, immediately remove the contaminated clothing and use the safety shower, if available. Make sure to inform your instructor about the incident. Do not leave your coats and back packs on the bench. No headphones and Walkman are allowed in the laboratory as they interfere with the ability to hear.

Fire

In case of fire or an accident, inform your instructor at once. Note the location of fire extinguishers and, if available, safety showers and safety blankets as soon as you enter the laboratory so that you may use them if needed. Never perform an unauthorized experiment in the laboratory.

I.8 Report format

Laboratory reports must be submitted on the same day, including date, title of the experiment, name of the student assigned by the instructor, etc. Late laboratory reports will be penalized accordingly.

Introduction

Briefly discuss the principle(s) of the experiment and give pertinent chemical equations.

Procedures

Cite the procedures from the lab manual or handout. You should also write the entire synthesis and be sure to include any deviations or modifications to the procedures. Your procedures should also identify any instrumentation used (make and model), and how samples were prepared, along with the conditions under which the samples were run.

Results and discussion

Break your results into the following categories (if applicable):
- Synthesis: describe the reaction briefly (color changes, precipitates, yields, etc). Indicate any problems or interesting aspects of the synthesis.
- Characterization: describe any type of spectroscopic results (IR, UV-vis, NMR, magnetic susceptibility, etc.); use as many figures as possible.
- Analysis: show pertinent analytical data (i.e. titration curve) in table and/or graphical format. Indicate any equations you have used in calculating an answer. If applicable, compare with the known literature values (cite your references!) and note any differences. If these are significant, discuss the reason(s) behind the observed difference.

I.9 Regulations

1. Laboratory coat and safety glasses

Note: No student is permitted to work in laboratory without coat and safety glasses. All students must wear laboratory coat and safety eyeglasses (or their own eyeglasses) at all times in the laboratory. Much of the danger of an injury to the eyes from spattering reagents as the result of accident (e.g. either to you or your neighbor) can be avoided by this precaution.

2. Laboratory tools and notebook

The keeping of a proper laboratory notebook is an essential part of all scientific work, and students should aim at keeping a book that is clear, tidy and original. Notes should be written directly in a properly bound notebook and should be dated. Each glassware should be chosen carefully according to the corresponding usage.

3. Notebook preparation

1. Take a fresh page for each experiment, and give the experiment a name, at the top of the page, e.g. Preparation of cyclohexanone.
2. Describe briefly, but accurately what you did (there is no point in writing the manual, especially if your experiment went according to plan), and what apparatus you used (there is no need to draw it). Weights of the reagents should be given.
3. Any particular observations such as color change should be recorded, as should unexpected observation - this is often the way the major scientific advances begin.
4. Include, in a prominent position, an equation to describe the reaction(s) you are reporting and formulate your interpretation of the reaction pathway (mechanism) involved.
5. The yield (weight and %) of the crude and purified products as well as the physical properties of all products should be reported.

4. Samples

Samples of all the reaction products you make should be kept. Special bottles are provided for this purpose and they should be labeled, showing your name, date, along with the name, B.P. or M.P. of the compound. Your final laboratory mark includes an assessment of these samples.

5. Calculations

Note 1): The calculation of the % yield: The percentage yield for any reaction represents the extent, expressed in percent, to which the reactants were converted into the isolated product. It can be calculated as follows:

$$\% \text{ yield} = \frac{\text{actual yield}}{\text{theoretical yield}} \times 100$$

Where, the actual yield is the weight of the product actually isolated, and the theoretical yield is the weight of the product that would be formed from the starting materials if the reaction proceeded to 100%, as indicated by the balanced equation. For example, let us calculate the % yield of the ester (ethyl acetate) if 33.6 g were isolated from the reaction of 35.0 g of acetic acid with 75.0 g of ethyl alcohol. The balanced equation is:

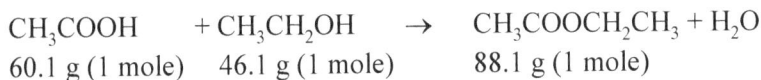

$$CH_3COOH \quad + CH_3CH_2OH \quad \rightarrow \quad CH_3COOCH_2CH_3 + H_2O$$

60.1 g (1 mole)   46.1 g (1 mole)   88.1 g (1 mole)

Theoretically, 60.1 g (1 mole) of CH_3CO_2H reacts with 46.1 (1 mole) of C_2H_5OH to yield 88.1 g (1 mole) of $H_3CO_2C_2H_5$. Hence, the theoretical yield of $CH_3CO_2C_2H_5$ from 35.0 g of CH_3CO_2H and 75.0 g of C_2H_5OH is:

$$\frac{35.0}{60.1} \quad x \quad 88.1 \quad = 51.3 \text{ g} \quad \text{and} \quad \frac{75.0}{46.1} \quad x \quad 88.1 \quad = 143.3 \text{ g}$$

In other words, the C_2H_5OH is present in theoretical excess, and CH_3CO_2H is the limiting in determining the theoretical yield. No matter how great an excess of C_2H_5OH added, the maximum yield of ethyl acetate that can be obtained from 35.0 g of CH_3CO_2H is 51.3 g.

Thus , the % yield is = $\dfrac{33.6}{51.3}$ x 100 % = 65.5%

Esterification is an equilibrium reaction, and the addition of an excess of the 'cheaper reagent' ethanol, through the mass action effect, produced per unit weight of the more costly reactant, acetic acid.

Note 2): The calculation of the number of moles: You may prefer to solve problems by calculating the number of moles of the limiting reactant used and the product isolated. Thus, in the synthesis of ethyl acetate, we see from the equation that, for every 1 mole of acetic acid, which reacts, 1 mole of ethyl acetate is formed. However, with 35.0 g of acetic acid, the theoretical yield of ethyl acetate is, therefore, 0.582 mole of ethyl acetate.

$$\frac{35.0 \text{ g}}{60.1 \text{ g/mole}} \quad = \quad \textbf{0.582 mole of acetic acid}$$

The actual yield of ethyl acetate is

$$\frac{33.6\ g}{88.1\ g/mole} = \textbf{0.381 mole of ethyl acetate}$$

The % yield is, therefore, calculated as

$$\frac{0.381\ mole}{0.582\ mole} \quad \text{x} \quad 100\% \quad = \quad \textbf{0.582 mole of acetic acid}$$

I.10 Techniques

In this part, the student can understand and conduct the following:

1. Laboratory safety: eye safety, fires, the hazard of organic solvents, waste solvent disposal, dispensing reagents, food in the laboratory and first aid.
2. Melting point determination
3. Boiling point determination (theory of simple and fractional distillation and semi-micro), azeotropic solution, separation of compounds through simple distillation, steam distillation: theory and apparatus, separation of organic compounds, etc.
4. Recrystallization (purification of solid organic compounds, choosing a solvent, gravity filtration, Buchner filtration, decolorization, etc)
5. Extraction: separation of a mixture
6. Sublimation: theory, apparatus and uses

I.11 Qualitative analysis

After the following qualitative analysis, the student can proceed to the experimental preparation:

1. Introduction to the qualitative analysis of compounds
2. Timetable of laboratory sessions and the purpose of the experiment.
3. Laboratory safety: eyes, fires, hazards of solvents, waste solvent disposal, dispensing reagents, food in the laboratory and first aid.
4. Along with the theoretical part, the student should conclude and summarize all knowledge about the primary tests for the classifications of inorganic, organic compounds and functional

groups and reactions as well.

5. The student should be able to differentiate between simple inorganic or organic compounds with simple reactions and tests.

6. The student should be able to understand and generate simple schemes in the sequence of experiments.

7. The experiments selected for study can performed as follows:

Inorganic compounds
a) Classification of acid radicals
b) Identification of basic radicals
c) Mixture separation
d) Volumetric analysis
e) Gravimetric analysis

Organic compounds
a) Physical characters
b) Dry heat
c) Solubility
d) Action of $NaHCO_3$ solution
e) Action of 30% NaOH solution
f) Action of soda-lime
g) Action of $FeCl_3$
h) Action of conc. H_2SO_4 (cold/hot)
i) Determination of constant (B.P./M.P.)
j) Detection of elements; N, S, X, etc.

8. The classification of experiments selected as typical chemistry textbook:
   a) Know how to distinguish different classes from each other.
   b) Know how to distinguish different class members from each other.
   c) Know how the physical properties for different members of different classes vary with their physical characteristics and chemical characteristics.
   d) Know how to prepare different inorganic and organic compounds in the laboratory to confirm the properties to be used as indications for particular compounds.

9. The most important functional groups among organic compounds are:
   a) Alcohols
   b) Aldehydes and ketones

c) Carboxylic acids
d) Salts of carboxylic acids
e) Amides
f) Phenols
g) Halogenated and fatty acids
h) Amines and amine salts
i) Esters
j) Anilides
k) Aromatic hydrocarbons
l) Halogenated hydrocarbons
m) Carbohydrates
n) Amino acids

I.12 Synthesis

In this part of advanced experimental preparation, the student can proceed with pharmaceutical chemistry courses by by applying most of the chemical reactions used in organic chemistry, as mentioned below:

A. Single-step synthesis

Note: The students should learn how to apply the previously mentioned techniques. Calculate the yield percent (%), the weight of a compound via mole fraction value and know the description of the glassware in the following:

1) Preparation of cyclohexene
2) Preparation of n -butyl bromide
3) Preparation of cyclohexanone
4) Preparation of n -amylacetate
5) Preparation of dibenzalacetone
6) Preparation of aspirin

B. Multi-step synthesis

1) Preparation of 3- aminobenzoic acid
   a) Preparation of methyl benzoate
   b) Nitration of methyl benzoate and recrystallization
   c) Preparation of 3- aminobenzoic acid
2) Preparation of p-chlorotoluene: Sandmayer reaction, diazotization
3) Preparation of benzocaine:
   a) Preparation of N-acetyl-p-toluidine: N-acetylation

b) Preparation of p-acetamidobenzoic acid: oxidation of aromatic methyl group

c) Preparation of p-aminobenzoic acid: hydrolysis of amide group and reflux

d) Preparation of benzocaine (ethyl p-aminobenzoate) esterification and reflux

C. Heterocyclic synthesis

1) Preparation of quinoline (Skraup synthesis)
2) Preparation of barbituric acid: cyclization reaction
3) Preparation of benzimidazoles
4) Preparation of tetrahydrocarbazole
5) Preparation of benzaoxazoles
6) Preparation of 5, 5-duiphenylhydrartoin (Phenytoin)

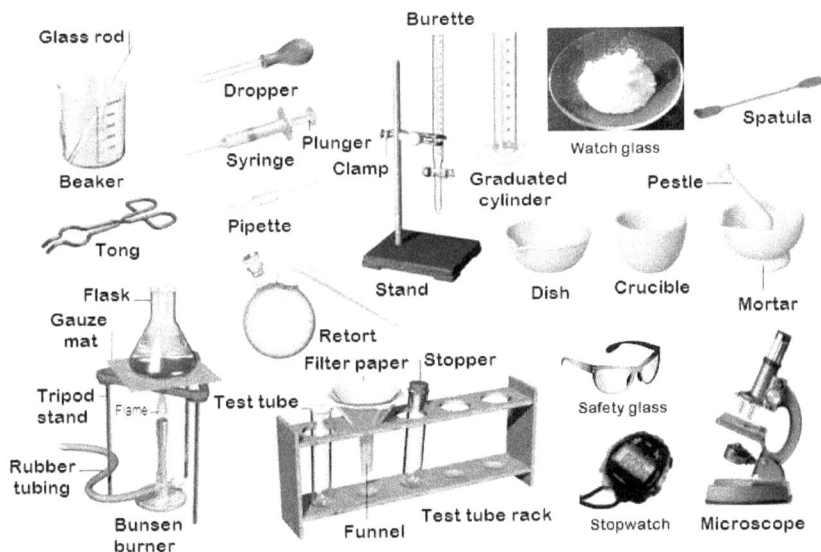

Chapter II

Qualitative Inorganic Chemical Analysis

II.1 Introduction

The students are familiar with most acids and bases before they start a formal course in chemistry. When ionic compounds dissolve in water, the ions will separate from each other and they mix with the water molecules. A solution of sodium carbonate in water consists of simply sodium ions, Na^+, and carbonate ions, CO_3^{2-}, mixed with water molecules. If sodium chloride is dissolved in water, the solution contains sodium ions Na^+, and chloride ions Cl^-. When any compounds or the salts are able to conduct electricity when dissolved in water, they called electrolytes. Solutions of hydrogen chloride and other hydrogen halides in water share certain properties with the aqueous solutions of a number of other substances. A solution of hydrogen chloride in water consists of simply hydronium ions, H_3O^+, and chloride ions, Cl^-, mixed with water molecules. These properties have long been used to identify these substances as acids. One of the main important characteristic properties of aqueous solution of acids results from the fact that, like a solution of hydrogen chloride in water, they all contain the hydronium ion. The other property is that acids have the ability to react with bases and salts such as sodium carbonate to produce carbon dioxide. As well acids can react with metals such as zinc or magnesium, producing hydrogen gas and the salts. From common knowledge, for instance, that acetic acid is an acid ingredient in vinegar, hydrochloric acid is in the digestive juices of stomach, and citric acid present in citrus fruits, such as, lemon, orange …etc. Many people recognize the strong pungent odor of the base ammonia, which is commonly used in cleaners as disinfectant and in water solution. The acids and bases have been defined by chemists as: an acid is a substance whose water solution has a sour taste and a base is a substance whose water solution has a bitter taste. Although these definitions have important practical value, several principles are presented to illustrate chemical behavior of such substances.

II.2 Acid and base theories

II.2.1 Arrhenius principle

In 1887 Svante Arrhenius defined the acids as those substances that dissolve in water to produce hydrogen ions in the media (for examples; HCl, $HClO_4$, HNO_3, H_2SO_4, H_3PO_4, HCO_2H, CH_3CO_2H) and bases as those substances that dissolve to give hydroxide ions in the media (for examples; KOH, NH_3, $Ba(OH)_2$, $Al(OH)_3$, $NaOH$).

II.2.2 Bronsted-Lowry principle

In 1923, Brönsted (Denmark Scientist) and Lowry (England Scientist) postulated that the acids are those substances which donate a proton and the bases are those substances which accept protons. By this definition, a great variety of chemical reactions and chemical properties can be correlated involved the chemical reactions that take place in solvents such as water. By considering the Brönsted-Lowry definitions for acids and bases, the proton donating compounds are acids and proton accepting compounds are bases. In water solution, compounds such as; HNO_3, HCO_2H and HCl are acids and NH_3 is considered to be as a base. They represent the anion, for example, NO_3^-, HCO_2^- and Cl^-. The following equations illustrate their reactions:

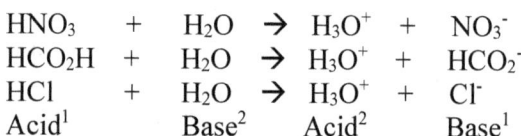

$$HNO_3 \quad + \quad H_2O \quad \rightarrow \quad H_3O^+ \quad + \quad NO_3^-$$
$$HCO_2H \quad + \quad H_2O \quad \rightarrow \quad H_3O^+ \quad + \quad HCO_2^-$$
$$HCl \quad + \quad H_2O \quad \rightarrow \quad H_3O^+ \quad + \quad Cl^-$$
$$Acid^1 \qquad Base^2 \qquad Acid^2 \qquad Base^1$$

$Acid^1$ is the strong acid, $base^2$ is weak base, $base^1$ is the conjugate base of $acid^1$; $acid^2$ is the conjugate acid of $base^1$.

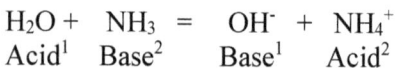

$$H_2O + \quad NH_3 \quad = \quad OH^- \quad + \quad NH_4^+$$
$$Acid^1 \quad Base^2 \qquad Base^1 \qquad Acid^2$$

Water acts as $acid^1$, and ammonia acts as a $base^1$, $base^2$ is the conjugate base of $acid^1$; $acid^2$ is the conjugate acid of $base^1$. In case of polyprotic acids, they are capable of donating more than one proton in more than one step, the acids, such as; H_2SO_4 and H_3PO_4 can donate two and three protons, respectively.

II.2.3 Lewis principle

In 1916, Lewis postulated that the acid is a substance which accepted an electron pair and the base is a substance which donated an electron pair. The following two examples illustrate the Lewis acid- base reactions:

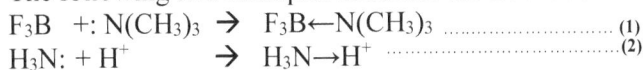

F_3B +: $N(CH_3)_3$ → F_3B←$N(CH_3)_3$ .... (1)

H_3N: + H^+ → H_3N→H^+ ....(2)

In the equation (1), F_3B represents Lewis acid and $N(CH_3)_3$ is Lewis base, but in the equation (2), H_3N is Lewis base and H^+ is Lewis acid. Many other chemical compounds can behave as acids and many behave as bases as well.

II.3 Salts

Salts are defined as "the product formed from the neutralization reactions of ionic compounds". They formed mainly from the reaction of acids and bases. These neutral components of the salts are electrically charged such as chlorides Cl^- in NaCl, as well organic ions, $CH_3CO_2^-$ in certain salts. Examples for the formation of salts from the reaction of acids and bases can be illustrated as followed,

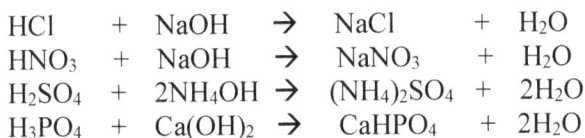

HCl + NaOH → NaCl + H_2O

HNO_3 + NaOH → $NaNO_3$ + H_2O

H_2SO_4 + $2NH_4OH$ → $(NH_4)_2SO_4$ + $2H_2O$

H_3PO_4 + $Ca(OH)_2$ → $CaHPO_4$ + $2H_2O$

II.4 Classification of acid and basic radicals

II.4.1 Preface

Some of acid radicals were classified into three groups to simplify their chemistry and their identification, as well to study their properties. The following table shows the common salts which are used in the identification of acid radicals.

The classification of the acid radicals was based on the strength and the stability of the acids. The strong acid replaces weak acid in its salts, and the most stable acid replaces the unstable acids. For this reason, the acid radicals were classified due to their reactions with hydrochloric and concentrated sulphuric acids.

Common salts

Salt	Anions	Cations	Formula
Silver nitrate	NO_3^-	Ag^+	$AgNO_3$
Aluminum chloride	Cl^-	Al^{3+}	$AlCl_3$
Aluminum nitrate	NO_3^-	Al^{3+}	$Al(NO_3)_3$
Barium chloride	Cl^-	Ba^{2+}	$BaCl_2$
Barium nitrate	NO_3^-	Ba^{2+}	$Ba(NO_3)_2$
Bismuth nitrate	NO_3^-	Bi^{3+}	$Bi(NO_3)_3$
Calcium chloride	Cl^-	Ca^{2+}	$CaCl_2$
Calcium bromide	Br^-	Ca^{2+}	$CaBr_2$
Calcium nitrate	NO_3^-	Ca^{2+}	$Ca(NO_3)_2$
Cadmium nitrate	NO_3^-	Cd^{2+}	$Cd(NO_3)_2$
Chromium(III) chloride	Cl^-	Cr^{3+}	$CrCl_3$
Copper sulphate	SO_4^{2-}	Cu^{2+}	$CuSO_4$
Iron(II) sulphate	SO_4^{2-}	Fe^{2+}	$FeSO_4$
Mercury(II) nitrate	NO_3^-	Hg^{2+}	$Hg(NO_3)_2$
Mercury(I) nitrate	NO_3^-	Hg_2^{2+}	$Hg_2(NO_3)_2$
Potassium chloride	Cl^-	K^+	KCl
Potassium bromide	Br^-	K^+	KBr
Potassium iodide	I^-	K^+	KI
Potassium nitrite	NO_2^-	K^+	KNO_2
Potassium nitrate	NO_3^-	K^+	KNO_3
Potassium oxalate	$C_2O_4^{2-}$	K^+	$K_2C_2O_4$
Potassium chromate	CrO_4^{2-}	K^+	K_2CrO_4
Magnesium chloride	Cl^-	Mg^{2+}	$MgCl_2$
Magnesium nitrate	NO_3^-	Mg^{2+}	$Mg(NO_3)_2$
Manganese chloride	Cl^-	Mn^{2+}	$MnCl_2$
Sodium Chloride	Cl^-	Na^+	$NaCl$
Sodium bromide	Br^-	Na^+	$NaBr$
Sodium nitrite	NO_2^-	Na^+	$NaNO_2$
Sodium nitrate	NO_3^-	Na^+	$NaNO_3$
Sodium iodide	I^-	Na^+	NaI
Sodium carbonate/ bicarbonate	CO_3^{2-}/ HCO_3^-	Na^+	$Na_xCO_3/NaHCO_3$
Sodium acetate	$CH_3CO_2^-$	Na^+	$NaCH_3CO_2$
Sodium sulphate	SO_4^{2-}	Na^+	Na_2SO_4
Sodium sulphite	SO_3^{2-}	Na^+	Na_2SO_3
Sodium thiosulphate	$S_2O_3^{2-}$	Na^+	$Na_2S_2O_3$
Sodium sulphide	S^{2-}	Na^+	Na_2S

Sodium phosphate	PO_4^{3-}	Na^+	Na_3PO_4
Sodium borate	$B_4O_7^{2-}$	Na^+	$Na_2B_4O_7$
Sodium dihyd. Phosphate	PO_4^{3-}	Na^+	NaH_2PO_4
Ammonium chloride	Cl^-	NH_4^+	NH_4Cl
Ammonium nitrate	NO_3^-	NH_4^+	NH_4NO_3
Ammonium carbonate	CO_3^{2-}	NH_4^+	$(NH_4)_2CO_3$
Ammonium phosphate	PO_4^{3-}	NH_4^+	$(NH_4)_3PO_4$
Lead acetate	$CH_3CO_2^-$	Pb^{2+}	$Pb(CH_3CO_2)_2$
Lead nitrate	NO_3^-	Pb^{2+}	$Pb(NO_3)_2$
Zinc nitrate	NO_3^-	Zn^{2+}	$Zn(NO_3)_2$

II.4.2 Reaction of acid radicals with dilute hydrochloric acid

This group known as hydrochloric acid group, its salts react with dilute HCl and different gases of different colors and odors will evolve. The salts are; carbonate (CO_3^{2-}), bicarbonate (HCO_3^-), nitrite (NO_2^-), sulphide(S^{2-}), sulphite (SO_3^{2-}) and thiosulphate ($S_2O_3^{2-}$).

Identification tests

To a small amount of the solid salt in a test tube, add few drops of dil. HCl. The following table shows the obtained gases and their analogous acid radicals.

Test	Observation	Gases	Acid radicals
Solid salt +HCl	Colorless gas is evolved with effervescence; a gas (CO_2) is odorless and produces a turbidity when passed into lime water forming calcium carbonate: $CO_2 + Ca(OH)_2 \rightarrow CaCO_3 + H_2O$	CO_2	Carbonate or Bicarbonate
Solid salt +HCl	Nitrous fumes (NO_2) evolved at the top of the tube; recognized by reddish brown vapors and a pale green solution; $2NaNO_2 + 2HCl \rightarrow 2NaCl + 2NO_2 + H_2$	NO_2	Nitrite
Solid salt +HCl	Colorless gas evolved; odor of rotten eggs (H_2S); blackens a filter paper moistened with	H_2S	Sulphide

	$Pb(CH_3COO)_2$,; $K_2S + 2HCl \rightarrow 2KCl + H_2S$ $H_2S + Pb(CH_3COO)_2 \rightarrow$ $PbS(Black) + 2CH_3COOH$		
Solid salt +HCl	Colorless gas evolved with suffocating odor; turns a filter paper moistened with potassium dichromate to green indicating the reduction of Cr^{6+} to Cr^{3+}.	SO_2	Sulphite
Solid salt +HCl	Same as sulphate; in addition to deposit of sulfur at the bottom of the test tube	SO_2	Thiosulphate

Confirmation tests

The confirmation tests for the acid radicals which react with dilute hydrochloric acid are summarized in the following table:

Magnesium sulphate solution with salt solution

CO_3^{-}	HCO_3^-	NO_2^-	S^{--}	SO_3^{--}	$S_2O_3^{--}$
White ppt of magnesium carbonate; $Na_2CO_3 + MgSO_4 \rightarrow$ $MgCO_3 + Na_2SO_4$	Same observation; but on heating	—	—	—	—

Mercury (II) chloride solution with salt solution

CO_3^{-}	HCO_3^-	NO_2^-	S^{--}	SO_3^{--}	$S_2O_3^{--}$
Reddish brown ppt of mercury(II) carbonate $Na_2CO_3 + HgCl_2 \rightarrow HgCO_3$ $+ 2NaCl$	Same observation; but on heating	—	—	—	—

Silver nitrate solution with salt solution

CO_3^{-}	HCO_3^-	NO_2^-	S^{--}	SO_3^{--}	$S_2O_3^{--}$
White ppt. of silver carbonate; $Na_2CO_3 +$ $2AgNO_3 \rightarrow$ $Ag_2CO_3 +$	Same as carbonate but on heating.	White ppt. of silver nitrite soluble on	Black ppt. of silver sulphide, soluble in hot	White ppt. of silver sulphate soluble in	White ppt. of silver thiosulphate turns to yellow then brown ppt

2NaNO$_3$		heating; NaNO$_2$ + AgNO$_3$ → AgNO$_2$↓ + NaNO$_3$	HNO$_3$; K$_2$S + AgNO$_3$ → Ag$_2$S +2KNO$_3$	excess of the reagent; Na$_2$SO$_3$ + 2AgNO$_3$ → Ag$_2$SO$_3$ + 2NaNO$_3$	in excess of reagent; Na$_2$S$_2$O$_3$ + 2AgNO$_3$ → Ag$_2$S$_2$O$_3$ + 2NaNO$_3$

Salt solution with lead acetate solution

CO$_3^{2-}$	HCO$_3^-$	NO$_2^-$	S^{2-}	SO$_3^{2-}$	S$_2$O$_3^{2-}$
White ppt. of lead carbonate Na$_2$CO$_3$ + Pb(CH$_3$CO$_2$)$_2$ → PbCO$_3$ + 2CH$_3$CO$_2$Na	Same as carbonate but on heating		Black ppt. of lead sulphide Na$_2$S + Pb(CH$_3$CO$_2$)$_2$ → PbS + 2CH$_3$CO$_2$Na	White ppt. of lead sulphate turns to black on heating Na$_2$SO$_3$ + Pb(CH$_3$CO$_2$)$_2$ → PbSO$_3$ + 2CH$_3$CO$_2$Na	White ppt. of lead thiosulphate turns to yellow then brown ppt; Na$_2$S$_2$O$_3$ + Pb(CH$_3$CO$_2$)$_2$ → PbS$_2$O$_3$+ 2CH$_3$CO$_2$Na

Acidified[*] Potassium permanganate solution with salt solution

CO$_3^{2-}$	HCO$_3^-$	NO$_2^-$	S^{2-}	SO$_3^{2-}$	S$_2$O$_3^{2-}$
—	—	The color of KMnO$_4$ disappeared, due to the oxidation-reduction process	—	—	The color of KMnO$_4$ disappeared, due to the oxidation-reduction process

Iodine solution with salt solution

CO$_3^{2-}$	HCO$_3^-$	NO$_2^-$	S^{2-}	SO$_3^{2-}$	S$_2$O$_3^{2-}$
—	—	—	—	The color of KMnO$_4$ disappeared, due to the oxidation-reduction process;	The color of KMnO$_4$ disappeared, due to the oxidation-reduction process;

				$Na_2SO_3 + I_2 + H_2O$ $\rightarrow Na_2SO_4 + 2HI$	$Na_2S_2O_3 + I_2 \rightarrow$ $2NaI + Na_2S_4O_6$

Acidified potassium dichromate (with dilute H_2SO_4) with salt solutions

CO_3^{2-}	HCO_3^-	NO_2^-	S^{2-}	SO_3^{2-}	$S_2O_3^{2-}$
—	—	—	—	Greenish color, due to the oxidation-reduction process	Greenish color, due to the oxidation-reduction process

Potassium iodide with salt solutions followed by dil. sulphuric acid

CO_3^{2-}	HCO_3^-	NO_2^-	S^{2-}	SO_3^{2-}	$S_2O_3^{2-}$
—	—	The solution colored with faint brown color due to separated crystals of iodine; $2KI + 2 KNO_2 + H_2SO_4 \rightarrow 2K_2SO_4 + 2NO + I_2\downarrow + 2H_2O$	—	Greenish color, due to the oxidation-reduction process	Greenish color, due to the oxidation-reduction process

Barium chloride solution with salt solution

CO_3^{2-}	HCO_3^-	NO_2^-	S^{2-}	SO_3^{2-}	$S_2O_3^{2-}$
White ppt of barium carbonate; soluble in dilute mineral acids; $Na_2CO_3 + BaCl_2 \rightarrow BaCO_3\downarrow + 2NaCl$	Same as carbonate but on heating	—	—	White ppt of barium sulphate; soluble in dil.HCl; $Na_2SO_3 + BaCl_2 \rightarrow BaSO_3\downarrow + 2NaCl$	—

Ferrous sulphate and dil. sulphuric acid with salt solution

CO_3^{2-}	HCO_3^-	NO_2^-	S^{2-}	SO_3^{2-}	$S_2O_3^{2-}$
—	—	Brown color	—	—	—

Ferric chloride solution with salt solution

CO_3^{2-}	HCO_3^{2-}	NO_2^-	S^{2-}	SO_3^{2-}	$S_2O_3^{2-}$
—	—	—	—	—	The color of the solution becomes violet, then disappeared after long time

Potassium thiocyanate solution with salt solution

CO_3^{2-}	HCO_3^-	NO_2^-	S^{2-}	SO_3^{2-}	$S_2O_3^{2-}$
—	—	Deep red color; disappeared on heating; $NaNO_2 + KSCN \rightarrow$ $NaSCN + KNO_2$	—	—	—

Sodium nitroprosside solution with salt solution

CO_3^{2-}	HCO_3^-	NO_2^-	S^{2-}	SO_3^{2-}	$S_2O_3^{2-}$
—	—	—	A addition of the reagent to the sodium sulphide (in basic media); the solution takes purple color due to the formation of sodium nitroprosside sulphide; $Na_2S + Na_2[Fe(CN)_5NO] \rightarrow$ $Na_4[Fe(CN)_5ONS]$	—	—

II.4.3 Reaction of acid radicals with concentrated sulphuric acid

The salts of this class react with conc. Sulphuric acid and different gases of different   colors and odors were evolved. It contains four salts which are; Chloride (Cl^-), Bromide (Br^-), Iodide (I^-) and nitrate (NO_3^-)

Identification tests

To a small amount of the solid salt in a test tube, add few drops of conc. H_2SO_4. The following table shows the obtained gases and their analogous acid radicals.

Test	Observation	Gases	Acid radicals
Solid salt +	Colorless hydrogen chloride vapor is evolved; white fumes	HCl	Chlorides

conc. H_2SO_4	were formed when passed through ammonium hydroxide solution; $2KCl + H_2SO_4 \rightarrow K_2SO_4 + 2HCl$ $HCl + NH_4OH \rightarrow NH_4Cl + H_2O$		
Solid salt + conc. H_2SO_4	Reddish brown vapors of bromine; turns the color of starch paper to yellow, and the mixture becomes orange or reddish-brown color; $2KBr + H_2SO_4 \rightarrow K_2SO_4 + 2HBr$ $2HBr + H_2SO_4 \rightarrow Br_2 + SO_2 + H_2O$	HBr/Br_2	Bromide
Solid salt + conc. H_2SO_4	Violet vapors of iodine (MnO_2 may add as catalyst); turns the color of starch to blue and the color of the mixture becomes brown; $2KI + H_2SO_4 \rightarrow K_2SO_4 + 2HI$ $2HI + H_{2S}O_4 \rightarrow I_2 + SO_2 + H_2O$	I_2	Iodide
Solid salt + conc. H_2SO_4 in presence of MnO_2	Brown vapors evolved at the top of the test tube; $2NaNO_3 + H_2SO_4 \rightarrow 2HNO_3 + Na_2SO_4^-$ $2HNO_3 + H_2SO_4 \rightarrow N_xO_y^{\#} + HSO_4^- + H_2O$ (# = NO, NO₂, or other nitrogen oxides)	NO_2^-	Nitrate

Confirmation tests

The confirmation tests for the acid radicals which react with conc. sulphuric acid are shown below:

Silver nitrate solution with salt solution

Cl⁻	Br⁻	I⁻	NO_3^-
White ppt of silver chloride insoluble in water; but soluble in ammonium	Canary yellow ppt of silver bromide, sparingly soluble in ammonium	Yellow ppt of silver iodide, insoluble in ammonium	—

hydroxide forming a silver complex; NaCl + AgNO$_3$ → AgCl + NaNO$_3$ AgCl + 2NH$_4$OH → [Ag(NH$_3$)$_2$]Cl	hydroxide; KBr + AgNO$_3$ → KNO$_3$ + AgBr AgBr + 2NH$_4$OH→ [Ag(NH$_3$)$_2$]Br$^-$ +2H$_2$O	hydroxide; KI +AgNO$_3$→ KNO$_3$ + AgI	

Lead acetate solution with salt solution

Cl$^-$	Br$^-$	I$^-$	NO$_3^-$
White ppt. of lead chloride soluble in hot water; 2NaCl + Pb(CH$_3$CO$_2$)$_2$ → PbCl$_2$ + 2CH$_3$CO$_2$Na	Yellowish white ppt. of lead bromide soluble in hot water; 2NaBr + Pb(CH$_3$CO$_2$)$_2$ → PbBr$_2$ + 2CH$_3$CO$_2$Na	Yellow ppt of lead iodide, soluble in hot water: 2KI + Pb(CH$_3$CO$_2$)$_2$ → PbI$_2$ + 2CH$_3$CO$_2$K	—

Ferrous sulphate solution with salt solution

Cl$^-$	Br$^-$	I$^-$	NO$_3^-$
—	—	—	Ferrous sulphate (in excess) added to nitrate solution followed by adding drops of conc. sulphuric acid on the wall of the test tube; a brown ring appeared. By stirring the tube, the mixture turns to brown color; 6FeSO$_4$ + 2HNO$_3$ + 3H$_2$SO$_4$ → 3Fe$_2$(SO$_4$)$_3$ + NO + 4H$_2$O [Fe(H$_2$O)$_6$]$^{3+}$ + NO → [Fe(H$_2$O)$_5$NO]SO$_4$ + H$_2$O

Manganese dioxide and conc. sulphuric acid with solid salt chlorine water and chloroform with salt solution

Cl$^-$	Br$^-$	I$^-$	NO$_3^-$
—	The addition of few drops of chlorine water to 2ml of bromide solution in test tube; reddish orange color appeared due to the entrance of chlorine instead of bromine and the separation of bromine; 2KBr + Cl$_2$ → 2KCl + Br$_2$	The addition of chlorine water to 2ml of iodide solution in test tube; brown color appeared due to the separation of iodine; 2KI + Cl$_2$ → 2KCl + I$_2$ Then, by addition of	—

Then, by addition of 2ml of chloroform with stirring, the lower layer ($CHCl_3$ layer) takes the red color due to the solubility of bromine in the organic layer.	2ml of $CHCl_3$ with stirring, the lower layer ($CHCl_3$ layer) takes the violet color due to the solubility of bromine in the organic layer.

Copper sulphate with salt solution

Cl^-	Br^-	I^-	NO_3^-
—	—	At addition of 2 ml from the salt solution to copper(II) sulphate. A brown ppt of copper(I) iodide appeared and iodine separated with brown color also and then the whole solution becomes brown; $4KI + 2CuSO_4 \rightarrow 2K_2SO_4 + Cu_2I_2 + I_2$	—

II.4.4 General group (Precipitation group)

The salts of this group did not react with both acids (dilute HCl and conc. H_2SO_4), but they have special tests. It includes; Sulphate (SO_4^{2-}), borate ($B_4O_7^{2-}$) and phosphate (PO_4^{3-}).

Barium chloride solution with salt solution

SO_4^{2-}	$B_4O_7^{2-}$	PO_4^{3-}
White ppt of Barium sulphate; insoluble in the acids; $Na_2SO_4 + BaCl_2 \rightarrow BaSO_4 + 2NaCl$	White ppt of Barium borate; soluble in dil. nitric acid; $Na_2B_4O_7 + BaCl_2 + 3H_2O \rightarrow Ba(BO_2)\downarrow + 2NaCl + 2H_3BO_3$	White ppt of barium phosphate; soluble in dil. Nitric acid; $Na_2HPO_4 + BaCl_2 \rightarrow BaHPO_4 + 2NaCl$

Silver nitrate solution with salt solution

SO_4^{2-}	$B_4O_7^{2-}$	PO_4^{3-}
White ppt of silver sulphate; $Na_2SO_4 +$	White ppt of silver borate; by heating turns to black and dissolve in acetic acid and also in ammonia solution;	Yellow ppt of silver phosphate; soluble in dilute nitric acid and in ammonia solution and directly form white ppt,

$2AgNO_3 \rightarrow$ $Ag_2SO_4 +$ $2NaNO_3$	$Na_2B_4O_7 + 2AgNO_3 +$ $3H_2O \rightarrow$ $2AgBO_2 + 2H_3BO_3 +$ $2NaNO_3$ $AgBO_2 + 3H_2O/\Delta \rightarrow$ $Ag_2O + 2H_3BO_3$	because due to the formation of HPO_4^{2-}; $Na_3PO_4 + 3AgNO_3 \rightarrow$ $Ag_3PO_4 + 3NaNO_3$ $Ag_3PO_4 + NH_4OH \rightarrow$ $Ag(NH_3)_2^+ + HPO_4^{2-}$ $+H_2O$

Lead acetate solution with salt solution

SO_4^{2-}	$B_4O_7^{2-}$	PO_4^{3-}
White ppt of lead sulphate; $Na_2SO_4 + Pb(CH_3CO_2)_2 \rightarrow$ $PbSO_4 + 2CH_3CO_2Na$	White ppt of lead borate $Na_2B_4O_7 + Pb(CH_3CO_2)_2 \rightarrow$ $PbB_4O_7 + 2CH_3CO_2Na$	—

Borax test with salt solution

SO_4^{2-}	$B_4O_7^{2-}$	PO_4^{3-}
—	To borate solution add one drop of phenol phthalein; a pink color will appear, this color disappeared by adding glycerol. On heating the pink color will return again.	—

Ferric chloride solution with salt solution

SO_4^{2-}	$B_4O_7^{2-}$	PO_4^{3-}
—	—	At addition of a reagent to the salt solution; yellowish white ppt of iron(III) phosphate formed; soluble in dilute mineral acids; $Na_2HPO_4 + FeCl_3 \rightarrow FePO_4 + 2NaCl + HCl$

II.5 Basic radicals

Many metal ions were classified into groups to simplify their identification and study their chemical properties. To separate them, the accuracy must be available during the additions and knowledge of the media in which the precipitation process occurred. Some of the compounds precipitated in acidic media while the other precipitated in basic media or the rest may precipitate in neutral media. The basic radicals which we are dealing with were classified into six groups, as summarized below.

Identification and confirmation tests

The following tables include the conclusion of the reaction between the basic radicals and the proper chemical reagents used. In a proper test tube contains 4-5 drops of the salt solution add a proper chemical reagent and note the obtained observations.

II.5.1 Group I

This group of basic radicals reacts with dilute hydrochloric acid, containing three cations; silver (Ag^+), Mercury(I) (Mercurous Hg_2^{2+}) and lead (Pb^{2+}).

Test	Observations	Cations,	Basic radical
Salt solution + dil. HCl	White ppt of silver chloride, insoluble in H_2O, but soluble in ammonia solution produces a complex of $[Ag(NH_3)_2]Cl$; $HCl + AgNO_3 \rightarrow AgCl \downarrow + HNO_3$ $AgCl + 2NH_3 \rightarrow [Ag(NH_3)_2]Cl$	$\circ Ag^+$	Silver
Salt solution + dil. HCl	White ppt of Mercury(I) chloride, $2HCl + Hg_2(NO_3)_2 \rightarrow Hg_2Cl_2 \downarrow + 2HNO_3$	Hg_2^{2+}	Mercury(I)
Salt solution + dil. HCl	White ppt of lead chloride, insoluble in cold water, but soluble in boiled water, $2HCl + Pb(NO_3)_2 \rightarrow PbCl_2 \downarrow + 2HNO_3$	Pb^{2+}	Lead

Sodium hydroxide solution with salt solution

Pb^{2+}	Hg_2^{2+}	$Ag+$
White ppt of lead hydroxide; insoluble in the excess of the reagent; $Pb(NO_3)_2 + 2NaOH \rightarrow$	Black ppt of Mercury(I) oxide Hg_2O insoluble in excess of the reagent and in ammonium hydroxide ; $Hg_2(NO_3)_2 + 2NaOH \rightarrow$	Brown ppt of silver oxide; soluble in dilute nitric acid and in ammonium hydroxide, but insoluble in excess of the reagent; $2AgNO_3 + 2NaOH \rightarrow$

Pb(OH)$_2$ + 2NaNO$_3$ Pb(OH)$_2$ + 2NaOH → Na$_2$PbO$_2$ + 2H$_2$O	Hg$_2$O↓ + 2NaCl + H$_2$O	Ag$_2$O↓ + 2NaNO$_3$ + H$_2$O. Ag$_2$O + 4NH$_4$OH → 2[Ag(NH$_3$)$_2$]OH

Ammonium hydroxide solution with salt solution

Pb^{2+}	Hg$_2$$^{2+}$	Ag+
White ppt of lead hydroxide , insoluble in the excess of the reagent; Pb(NO$_3$)$_2$ + 2NH$_4$OH → Pb(OH)$_2$ + 2NH$_4$NO$_3$	Black ppt , containing free mercury and mercury(II) ammine complex insoluble in the excess of the reagent; Hg$_2$(NO$_3$)$_2$ + 2NH$_4$OH → HgNH$_2$Cl↓ + Hg↓ + NH$_4$Cl + 2H$_2$O	White ppt of silver hydroxide, rapidly turns to brown due to the formation of silver oxide which soluble in excess of the reagent; Ag$_2$O + 4NH$_4$OH → 2[Ag(NH$_3$)$_2$]OH

Potassium chromate with salt solution

Pb^{2+}	Hg$_2$$^{2+}$	Ag+
Yellow ppt of lead chromate soluble in dilute nitric acid and in sodium hydroxide, but insoluble in ammonium hydroxide; Pb(NO$_3$)$_2$ + K$_2$CrO$_4$ → PbCrO$_4$↓ + 2KNO$_3$	Orange ppt of Mercury(I) chromate; K$_2$CrO$_4$ + Hg$_2$(NO$_3$)$_2$ → Hg$_2$CrO$_4$↓ + 2KNO$_3$	Brick red ppt of silver chromate soluble in dilute nitric acid and in ammonium hydroxide; 2AgNO$_3$ + K$_2$CrO$_4$ → Ag$_2$CrO$_4$↓ + 2KNO$_3$

Potassium iodide solution with salt solution

Pb^{2+}	Hg$_2$$^{2+}$	Ag+
Yellow ppt of lead iodide soluble in hot water gives colorless solution; but returns on cooling to precipitate again as plates; Pb(NO$_3$)$_2$ + 2KI → PbI$_2$↓ + 2KNO$_3$. The plates soluble in excess of the reagent forming a complex; K$_2$[Pb(I)$_4$]	Greenish yellow ppt of Mercury(I) iodide insoluble in excess of the reagent; 2KI + Hg$_2$(NO$_3$)$_2$ → Hg$_2$I$_2$↓ + 2KNO$_3$	Yellow ppt of silver iodide insoluble in acids and ammonium hydroxide; 2AgNO$_3$ + 2KI → 2AgI↓ + 2KNO$_3$

II.5.2 Group II

This group of basic radicals reacts with H_2S in presence of dil. HCl (acidic media). This includes two subgroups; (a) includes ions of; mercury(II) (mercuric, Hg^{2+}), cadmium (Cd^{2+}), bismuth (Bi^{3+}) and copper(II) (Cu^{2+}) ions and (b); includes salts of arsenic(III, V) (arsenous, As^{3+}, arsenic As^{5+}) , stannic (II, IV) (stannous, Sn^{2+}, stannic Sn^{4+}) and antimony(III, V) (antimonus Sb^{3+}, antimonic Sb^{5+}) .

Subgroup (a)

Test	Observation	Cation,	Basic radical
Salt solution + dil. HCl + H_2S	White ppt of mercury (II) sulphide according to the reaction; $Hg(NO_3)_2$ +H_2S → HgS↓ + $2HNO_3$ The formed ppt rapidly turns to yellow then brown and finally to black. Insoluble in potassium hydroxide and in hot dilute nitric acid and soluble in the aqua regia (3:1 parts of HCl:HNO_3)	Hg^{2+}	mercury (II)
Salt solution (dil.) + dil. HCl + H_2S	Yellow ppt of cadmium sulphide; insoluble in potassium hydroxide but soluble in hot dilute acids; $CdSO_4 + H_2S$ → CdS↓ + H_2SO_4	Cd^{2+}	Cadmium
Salt solution + dil. HCl + H_2S	Black-brown ppt of bismuth sulphide; insoluble in potassium hydroxide but soluble in hot dilute nitric acid; $2Bi(NO_3)_2 + 3H_2S$ → Bi_2S_3↓ + $6HNO_3$	Bi^{3+}	Bismuth
Salt solution + dil. HCl + H_2S	Black sulphide of copper (II) sulphide; insoluble in potassium hydroxide but soluble in hot dilute nitric acid; $CuSO_4 + H_2S$ → CuS↓ + H_2SO_4	Cu^{2+}	Copper

Sodium hydroxide solution with salt solution

Cu^{2+}	Bi^{3+}	Cd^{2+}	Hg^{2+}
Blue ppt of copper(II) hydroxide; on heating, the color turns to black due to the formation of copper(II) oxide; $Cu(NO_3)_2 + 2NaOH \rightarrow Cu(OH)_2\downarrow + 2NaNO_3$ $Cu(OH)_2 +$ heating $\rightarrow CuO\downarrow + H_2O$	White ppt of bismuth hydroxide; on heating, the color of the precipitate turns to yellow due to the formation of basic bismuth oxide; $Bi(NO_3)_3 + 3NaOH \rightarrow Bi(OH)_3\downarrow + 3NaNO_3$ $Bi(OH)_3 +$ heating $\rightarrow BiO(OH)\downarrow + H_2O$	White ppt of cadmium hydroxide; insoluble in excess of the reagent; $Cd(NO_3)_2 + 2NaOH\rightarrow Cd(OH)_2\downarrow + 2NaNO_3$	Brown ppt of mercury (II) hydroxide; immediately turns to yellow at addition of excess of the reagent due to the formation of yellow Mercury(II) oxide; $HgCl_2 + 2NaOH \rightarrow HgO\downarrow + 2NaCl + H_2O$

Ammonium hydroxide solution with salt solution

Cu^{2+}	Bi^{3+}	Cd^{2+}	Hg^{2+}
Pale blue ppt of copper (II) hydroxide; soluble in excess of the reagent; $2CuSO_4 + 2NH_4OH \rightarrow Cu(OH)_2\downarrow + (NH_4)_2SO_4$ $Cu(OH)_2 + (NH_4)_2SO_4 + 6NH_3 \rightarrow 2[Cu(NH_3)_4]SO_4 + H_2O$	White ppt of bismuth hydroxide; insoluble in excess of the reagent; $Bi(NO_3)_3 + 2NH_4OH \rightarrow Bi(OH)_3\downarrow + 2NH_4NO_3$	White ppt of cadmium hydroxide; soluble in excess of the reagent; $CdSO_4 + NH_4OH \rightarrow Cd(OH)_2 \downarrow + (NH_4)_2SO_4$	Black ppt of mercury (II) amino chloride; solution; $HgCl_2 + 3NH_4OH \rightarrow HgNH_2Cl \downarrow + NH_4Cl + H_2O$

Potassium iodide solution with salt solution

Cu^{2+}	Bi^{3+}	Cd^{2+}	Hg^{2+}
White ppt of copper(I) iodide, then the solution becomes brown due to the separation of iodine; $CuSO_4 + 2KI \rightarrow$ $CuI_2\downarrow + K_2SO_4$ $2CuI_2 \rightarrow Cu_2I_2 + I_2$	Dark brown ppt of bismuth iodide; soluble in excess of the reagent forming a yellow solution; $Bi(NO_3)_3 + 2KI \rightarrow$ $BiI_3\downarrow + 3KNO_3$ $BiI_3 + KI \rightarrow K[BiI_4]$	—	Red ppt of mercury (II) iodide; soluble in excess of the reagent forming colorless solution; $Hg(NO_3)_2 + 2KI$ $\rightarrow$ $HgI_2\downarrow + 2KNO_3$ $HgI_2 + 2KI \rightarrow$ $K_2[HgI_4]$

Potassium ferrocyanide solution with salt solution

Cu^{2+}	Bi^{3+}	Cd^{2+}	Hg^{2+}
Reddish brown ppt of copper (II) Ferrocyanide; soluble in ammonium hydroxide forming blue solution; $2Cu(NO_3)_2 + K_4[Fe(CN)_6] \rightarrow$ $Cu_2[Fe(CN)_6] + 4KNO_3$	—	White ppt of cadmium ferrocyanide $2Cd(NO_3)_2 + K_4[Fe(CN)_6] \rightarrow$ $Cd_2[Fe(CN)_6] + 4KNO_3$	—

Subgroup (b)

Test	Observation	Cation	Basic radical
Salt solution + dil HCl +H_2S	- Yellow ppt of arsenous sulphide; soluble in sodium hydroxide and ammonium sulphide, but insoluble in HCl; $2As(NO_3)_3 + 3H_2S \rightarrow As_2S_3\downarrow + 6HNO_3$. - On warming and in the presence of dil. HCl , the same precipitation of arsenic salt solution formed; $2As(NO_3)_5 + 5H_2S \rightarrow As_2S_5\downarrow + 10HNO_3$	As^{3+} As^{5+}	Arsenous, Arsenic

	Arsenic sulphide soluble potassium hydroxide, sodium hydroxide, ammonium sulphide and ammonium carbonate		
Salt solution + dil HCl +H_2S	- Brown ppt of stannous sulphide; soluble in ammonium sulphide and hot conc. HCl, but insoluble in ammonium carbonate; $2SnCl_2 + 2H_2S \rightarrow SnS\downarrow + 2HCl$ - Yellow ppt of stannic sulphide; soluble in potassium hydroxide, ammonium sulphide and hot conc. HCl, but insoluble in ammonium carbonate solution. $SnCl_4 + 2H_2S \rightarrow SnS_2\downarrow + 4HCl$	Sn^{2+} Sn^{4+}	Stannous, Stannic
Salt solution + dil HCl +H_2S	- Orange ppt of antimony (III) sulphide is formed; $2SbCl_3 + 3H_2S \rightarrow Sb_2S_3\downarrow + 6HCl$ - Reddish orange ppt of antimony (IV) is formed; $SbCl_5 + 5H_2S \rightarrow Sb_2S_5 + 10HCl$ - Both precipitates are soluble in potassium hydroxide, hot conc. HCl and ammonium carbonate, but insoluble in ammonium solution.	Sb^{3+} Sb^{5+}	Antimony(III), Antimony (V)

Sodium hydroxide solution with salt solution

Sb^{3+}, Sb^{5+}	Sn^{2+}, Sn^{4+}	As^{3+}, As^{5+}
White ppt of antimony(III) oxide; soluble in excess of the reagent forming sodium antimonite; $2SbCl_3 + 6NaOH \rightarrow$	- ☉White ppt of stannous hydroxide; soluble in the excess of the reagent forming sodium stannite; $SnCl_2 + 2NaOH \rightarrow Na_2SnO_2 +$	___

| $Sb_2O_3 + 6NaCl + 3H_2O$. No precipitation in case of Sb^{5+} but a solution contains antimony(V) ions as complex; $[Sb(OH)_6]^-$ | $2H_2O$ - White ppt of stannic hydroxide soluble in excess of the reagent forming sodium stannate; $SnCl_4 + 4NaOH \rightarrow Sn(OH)_4 + 4NaCl$ $Sn(OH)_4 + 2NaOH \rightarrow Na_2SnO_3 + 3H_2O$ | |

Silver nitrate solution with salt solution

Sb^{3+}, Sb^{5+}	Sn^{2+}, Sn^{4+}	As^{3+}, As^{5+}
Black ppt of silver antimonide; which decomposed in excess of the reagent to silver metal and antimony(III) oxide; $SbCl_3 + 3AgNO_3 \rightarrow Ag_3Sb\downarrow + 3HNO_3$ $4Ag_3Sb\downarrow + 12AgNO_3 + 6H_2O \rightarrow$ $24Ag + Sb_4O_6 + 12 HNO_3$. **This test called Hofmann Test**	____	Yellow ppt of silver arsenate (the arsenite solution must be neutral); soluble in ammonium hydroxide and dilute nitric acid; $Na_3AsO_3 + 3AgNO_3 \rightarrow Ag_3AsO_3 + 3NaNO_3$ While, a dark brown ppt of silver arsenate (the solution of arsenate must be neutral); $Na_3AsO_4 + 3AgNO_3 \rightarrow Ag_3AsO_4 + 3NaNO_3$

Magnesia mixture with salt solution

Sb^{3+}, Sb^{5+}	Sn^{2+}, Sn^{4+}	As^{3+}, As^{5+}
____	____	In neutral media, a white ppt of ammonium magnesium arsenate; $Na_2HAsO_4 + MgCl_2 + NH_3 \rightarrow Mg(NH_4)ASO_4 + 2NaCl$. No reaction (no ppt) For arsenous salts.

II.5.3 Group III

This group of basic radicals reacts with ammonium hydroxide in presence of ammonium chloride, containing; **aluminum (Al^{3+}), Iron (II, III) (Ferrous, Fe^{2+} and Ferric, Fe^{3+}), and chromium (III) (Cr^{3+}).**

Test	Observation	Cation	Basic radical
Salt solution +NH_4Cl +NH_4OH	Gelatinous White ppt of aluminum hydroxide; soluble in mineral acids and in sodium hydroxide; $AlCl_3 + 3NH_4OH \rightarrow Al(OH)_3\downarrow + 3NH_4Cl$ $Al(OH)_3 + NaOH \rightarrow NaAlO_2 + 2H_2O$	Al^{3+}	Aluminum(III)
	Greenish gray ppt of Cr(OH)3; soluble in mineral acids and in sodium hydroxide; $CrCl_3 + 3NH_4OH \rightarrow Cr(OH)_3 + 3NH_4Cl$ $Cr(OH)_3 + NaOH \rightarrow NaCrO_2 + 2H_2O$	Cr^{3+}	Chromium(III)
	- White ppt of iron(II) hydroxide; which turns to pale green ppt due to the oxidation by the air; $FeSO_4 + 2NH_4OH \rightarrow Fe(OH)_2\downarrow + (NH_4)_2SO_4$	Fe^{2+}	Iron(II)
	- Gelatinous brown ppt of iron(III) hydroxide; insoluble in excess of the sodium hydroxide; $FeCl_3 + 3NH_4OH \rightarrow Fe(OH)_3 + 3NH_4Cl$	Fe^{3+}	Iron(III)

Disodium hydrogen phosphate solution with salt solution

Fe^{2+}, Fe^{3+}	Cr^{3+}	Al^{3+}
White ppt, turns into yellow of iron(III) phosphate; insoluble in acetic acid but soluble in mineral acids; $FeCl_3 + 2Na_2HPO_4 \rightarrow FePO_4\downarrow + 3NaCl + NaH_2PO_4$	Greenish gray ppt of chromium(III) phosphate; soluble in mineral acids; $CrCl_3 + 2Na_2HPO_4 \rightarrow CrPO_4\downarrow + 3NaCl + NaH_2PO_4$	White ppt of aluminum phosphate; insoluble in acetic acid, but soluble in the mineral acids and NaOH and KOH; $Al(NO_3)_3 + 2Na_2HPO_4 \rightarrow AlPO_4 \downarrow + 3NaNO_3 + NaH_2PO_4$

General Practical Chemistry

Ammonium sulphide solution with salt solution

Fe^{2+}, Fe^{3+}	Cr^{3+}	Al^{3+}
For iron(II) salts, a black iron(II) sulphide ppt; soluble in hydrochloric acid librating H_2S gas; $FeSO_4 + (NH_4)_2S \rightarrow FeS\downarrow$ $+(NH_4)_2SO_4$ $FeS + 2HCl \rightarrow FeCl_2 + H_2S$ For iron(III) salts, a black ppt of iron(III) sulphide soluble in hydrochloric acid forming iron(III) chloride, sulfur and librating H_2S; $FeCl_3 + 3(NH_4)_2S \rightarrow Fe_2S_3\downarrow +$ $6NH_4Cl$	Chromium(III) sulphide; which decomposes by H_2O into chromium(III) hydroxide; $2CrCl_3 +$ $3(NH_4)_2S \rightarrow$ $Cr_2S_3 + 6NH4Cl$ $Cr_2S_3 + 6H_2O \rightarrow$ $2Cr(OH)_3\downarrow +$ $3H_2S$	White ppt of aluminum sulphide; which decomposed by H_2O into aluminum hydroxide; $2Al(NO_3)_3 +$ $3(NH_4)_2S \rightarrow$ $Al_2S_3\downarrow +$ $6NH_4NO_3$ $Al_2S_3 + 6H_2O \rightarrow$ $Al(OH)_3\downarrow +$ $3H_2S$

Sodium hydroxide solution with salt solution

Fe^{2+}, Fe^{3+}	Cr^{3+}	Al^{3+}
Pale green ppt of iron(II) hydroxide, meanwhile a gelatinous brown ppt of iron(III) hydroxide, insoluble in excess of the reagent, but soluble in mineral acids	Greenish gray ppt of chromium hydroxide soluble in the excess of the reagent and in mineral acids	Gelatinous white ppt of aluminum hydroxide soluble in excess of the reagent and in mineral acids

Potassium ferrocyanide solution with salt solution

Fe^{2+}, Fe^{3+}	Cr^{3+}	Al^{3+}
White ppt of iron(II) ferricyanide; which rapidly turns to blue ppt due to the oxidation into iron(III) ferricyanide; $2FeSO_4 + K_4[Fe(CN)_6] \rightarrow Fe_2[Fe(CN)_6]\downarrow +$ $2K_2SO_4$ - Blue ppt for iron(III) salts $2FeCl_3 + K_4[Fe(CN)_6] \rightarrow Fe_2[Fe(CN)_6]_3\downarrow +$ $2K_2SO_4$	—	—

Potassium ferricyanide solution with salt solution

Fe^{2+}, Fe^{3+}	Cr^{3+}	Al^{3+}
- Dark blue ppt of Ferrous ferricyanide;		

$3FeSO_4 + 2K_3[Fe(CN)_6] \rightarrow Fe_3[Fe(CN)_6]_2\downarrow + 3K_2SO_4$ - For iron(III) salts, the color of the solution becomes brown due to the formation of iron(III) ferricyanide; $FeCl_3 + K_3[Fe(CN)_6] \rightarrow 3KCl + Fe[Fe(CN)_6]\downarrow$	——	——

II.5.4 Group IV

This group of basic radicals reacts with sulphide ion in presence of NH_4Cl and NH_4OH (basic media), containing; manganese (II), cobalt (II), nickel (II) and zinc (II).

Test	Observation	Cation	Basic radical
Salt solution + NH_4Cl + NH_4OH + H_2S	Buff ppt of manganese sulphide; soluble in mineral acids and acetic acid; $MnCl_2 + H_2S \rightarrow MnS\downarrow + 2HCl$	Mn^{2+}	Manganese
Salt solution + NH_4Cl + NH_4OH+ H_2S	White ppt of zinc sulphide; soluble in mineral acids; $ZnSO_4 + H_2S \rightarrow ZnS\downarrow + H_2SO_4$	Zn^{2+}	Zinc
Salt solution + NH_4Cl + NH_4OH+ H_2S	Black ppt of cobalt sulphide; insoluble in hydrochloric acid but soluble in hot conc. nitric acid and aqua regia	Co^{2+}	Cobalt
Salt solution + NH_4Cl + NH_4OH + H_2S	Black ppt of nickel sulphide; insoluble in hydrochloric acid but soluble in hot conc. nitric acid and aqua regia	Ni^{2+}	Nickle

Sodium hydroxide solution with salt solution

Ni^{2+}	Co^{2+}	Zn^{2+}	Mn^{2+}
Green ppt of nickel(II) hydroxide; insoluble in the excess of the reagent, but soluble in NH_4OH giving blue solution; $Ni(NO_3)_2 +$ $2NaOH \rightarrow$ $Ni(OH)_2 +$ $2NaNO_3$	Blue ppt of basic cobalt(II) hydroxide; $Co(NO_3)_2 +$ $NaOH \rightarrow$ $Co(OH)(NO_3)\downarrow +$ $NaNO_3$. Then turns to pink color by the addition of the excess of the reagent forming $Co(OH)_2$. Addition of oxidizing agent to the mixture the color becomes black due to the formation of $Co(OH)_3$.	White ppt of zinc hydroxide; soluble in excess of the reagent forming sodium zincate; $ZnCl_2+2NaOH$ $\rightarrow Zn(OH)_2\downarrow$ $+ 2NaCl$ $Zn(OH)_2 +$ $2NaOH\rightarrow$ $Na_2ZnO_2 +$ $2H_2O$	White ppt of manganese(II) hydroxide; insoluble in excess of the reagent but changed into dark brown color due to the formation of basic manganese(III) oxide; $MnCl_2 + 2NaOH$ $\rightarrow Mn(OH)_2\downarrow +$ $2NaCl$ $2Mn(OH)_2 + O_2$ $\rightarrow 2MnO(OH)\downarrow$

Potassium ferrocyanide solution with salt solution

Ni^{2+}	Co^{2+}	Zn^{2+}	Mn^{2+}
—	—	White ppt of zinc ferrocyanide; soluble in excess of the reagent; $2ZnCl_2 + K_4[Fe(CN)_6] \rightarrow$ $Zn_2[Fe(CN)_6]\downarrow + 4KCl$	—

Dimethylglyoxime solution with salt solution

Ni^{2+}	Co^{2+}	Zn^{2+}	Mn^{2+}
Red ppt of bisdimethylglyoximatonickle(II) in the presence of ammonium hydroxide; $NiCl_2 + 2HDMG \rightarrow$ $[Ni(DMG)_2]\downarrow + 2HCl$	—	—	—

Ammonium thiocyanate solution with salt solution

Ni^{2+}	Co^{2+}	Zn^{2+}	Mn^{2+}
—	Blue solution of cobalt(II) thiocyanate; $CoCl_2 + 2NH_4SCN \rightarrow Co(SCN)_2 + 2NH_4Cl$	—	—

Disodium hydrogen phosphate solution with salt solution

Ni^{2+}	Co^{2+}	Zn^{2+}	Mn^{2+}
Pale green ppt of nickel(II) ammonium phosphate; $NiCl_2 + NH_4OH + Na_2HPO_4 \rightarrow Ni(NH_4)PO_4 + 2NaCl + H_2O$	Violet ppt of cobalt(II) ammonium phosphate; $CoCl_2 + NH_4OH + Na_2HPO_4 \rightarrow Co(NH_4)PO_4 + 2NaCl + H_2O$	☼White ppt of zinc ammonium phosphate (in the presence of NH_4Cl); $ZnCl_2 + Na_2HPO_4 + NH_4Cl \rightarrow Zn(NH_4)PO_4 + 2NaCl$	Buff ppt of manganese(II) ammonium phosphate; $MnCl_2 + NH_4OH + Na_2HPO_4 + H_2O \rightarrow Mn(NH_4)PO_4.7H_2O$

Lead dioxide and concentrated nitric acid with salt solution

Ni^{2+}	Co^{2+}	Zn^{2+}	Mn^{2+}
—	—	—	The two reagents were added to the manganese salt(except chloride), then boiled, the manganic acid is formed and gives the upper layer a violet color; $2MnSO_4 + 5PbO_2 + 6HNO_3 \rightarrow 2HMnO_4 + 3Pb(NO_3)_2 + 2PbSO_4 + 2H_2O$

Sodium bismuthate and concentrated nitric acid with salt solution

Ni^{2+}	Co^{2+}	Zn^{2+}	Mn^{2+}
—	—	—	Add conc. HNO_3. Then dilute the mixture, then add small amount of sodium bismuthate and watch the formed violet color; $2MnSO_4 + 5NaBiO_2 + 16HNO_3 \rightarrow 2HMnO_4 + 2Na_2SO_4 + 5Bi(NO_3)_3 + NaNO_3 + 7H_2O$

II.5.5 Group V

This group of basic radicals reacts with ammonium carbonate in the presence of NH_4Cl and NH_4OH, containing; calcium, strontium and barium ions.

Test	Observation	Cation	Basic radical
Salt solution + NH_4Cl + NH_4OH + $(NH_4)_2 CO_3$	White ppt of calcium carbonate soluble in acetic acid and mineral acids; $CaCl_2 + NH_4Cl + NH_4OH + (NH_4)_2CO_3 \rightarrow CaCO_3\downarrow + 2NH_4Cl$	Ca^{2+}	Calcium
Salt solution + NH_4Cl + NH_4OH + $(NH_4)_2 CO_3$	White ppt of strontium carbonate soluble in acetic acid and mineral acids; $SrCl_2 + NH_4Cl + NH_4OH + (NH_4)_2CO_3 \rightarrow SrCO_3\downarrow + 2NH_4Cl$	Sr^{2+}	Strontium
Salt solution + NH_4Cl + NH_4OH + $(NH_4)_2CO_3$	White ppt of barium carbonate soluble in mineral acids; $BaCl_2 + NH_4Cl + NH_4OH + (NH_4)_2CO_3 \rightarrow BaCO_3\downarrow + 2NH_4Cl$	Ba^{2+}	Barium

Calcium sulphate solution with salt solution

Ba^{2+}	Sr^{2+}	Ca^{2+}
Intensive white ppt of barium sulphate on cold is formed; $BaCl_2 + CaSO_4 \rightarrow BaSO_4\downarrow + CaCl_2$	White ppt of strontium sulphate on heating / boiling or in standing; $SrCl_2 + CaSO_4 \rightarrow SrSO_4\downarrow + CaCl_2$	___

Ammonium oxalate solution with salt solution

Ba^{2+}	Sr^{2+}	Ca^{2+}
White ppt of barium oxalate soluble in mineral acids ;	White ppt of strontium oxalate soluble in mineral acids ;	White ppt of calcium oxalate (from conc. solution) soluble in mineral acids but

$BaCl_2 + NH_4Cl +$ $(NH_4)_2C_2O_4 \rightarrow$ $BaC_2O_4 + 3NH_4Cl$	$SrCl_2 + NH_4Cl +$ $(NH_4)_2C_2O_4 \rightarrow$ $SrC_2O_4 + 3NH_4Cl$	insoluble in acetic acid; $Ca(NO_3)_2 + NH_4Cl +$ $(NH_4)_2C_2O_4 \rightarrow$ $CaC_2O_4 + 2NH_4NO_3$

Potassium chromate solution with salt solution

Ba^{2+}	Sr^{2+}	Ca^{2+}
Yellow ppt of $BaCrO_4$, insoluble in acetic acid; $BaCl_2 + K_2CrO_4$ $\rightarrow$ $BaCrO_4 + 2KCl$	Yellow ppt of $SrCrO_4$ (from conc. solutions or heating), partially soluble in acetic acid; $Sr(NO_3)_2 + K_2CrO_4 \rightarrow$ $SrCrO_4 + 2KNO_3$	Yellow ppt of calcium chromate (from very conc. solutions), soluble in mineral acids; $Ca(NO_3)_2 + K_2CrO_4 \rightarrow$ $CaCrO_4 + 2KNO_3$

Flame tests

Ba^{2+}	Sr^{2+}	Ca^{2+}
Apple green	Crimson (deep red)	Brick red(reddish-brown)

II.5.6 Group VI

No group reagent for the basic radicals of this group which contains; Ammonium, magnesium, sodium and potassium ions.

Test	Observation	Cation	Basic radical
Salt solution + NaOH	Libration of the ammonia gas (sometimes heating may be needed); $NH_4Cl + NaOH \rightarrow$ $NH_3\uparrow + NaCl + H_2O$	NH_4^+	Ammonium
Salt solution + Nessler test	Brown color of ammonium tetraiodomercurate(II); $K_2[Hg(I)_4] + 2NH_4Cl \rightarrow$ $(NH_4)_2[Hg(I)_4] + 2KCl$	NH_4^+	Ammonium
Salt solution + Nessler test	Brown ppt of magnesium tetraiodomercurate(II); $2K_2[Hg(I)_4] + 2Mg(NO_3)_2 \rightarrow$ $2Mg[Hg(I)_4] + 4KNO_3$	Mg^{2+}	Magnesium

Salt solution + Flame test	Golden yellow	Na^+	Sodium
Salt solution + Flame test	Violet	K^+	Potassium

II.6 Separation and identification of basic radical mixtures

In order to analyze the mixtures of basic radicals, the following information should be known:

1. The precipitation and precipitation medium.
2. The precipitation temperature conditions.
3. Complete precipitation should be tested before proceeding to any further steps in the separation. This technique must be performed by the instructor to avoid any interference between group/groups cations.
4. The classification of basic radicals.
5. The tools and the methods of experiments to be carried without exceptions.

Therefore, according to the previous classification of the basic radicals, the steps to be followed should be understand and organized as follows:

1. Group I: precipitated in acidic medium.
2. Group II: precipitated in acidic medium with H_2S as precipitant.
3. Group III: precipitated by NH_4OH in presence of NH_4Cl.
4. Group IV: precipitated by H_2S in presence NH_4OH and NH_4Cl.
5. Group V: precipitated by $(NH_4)_2CO_3$ in presence NH_4OH and NH_4Cl.
6. Group VI: can be precipitated by any of the previous group reagents.

The coming summary and tables include the conclusion of the reaction between the basic radicals and the proper chemical reagents used. The steps should be followed carefully to avoid any interference or not needed steps that lead to different results, and at the end you will find yourself with different element cations. All are misleading, because due to the similarity of most chemical reactions between the basic radicals.

You should consult the tests for each cation in each group for confirmation.

Note: If you are using filtrates, i.e. you have to keep the filtrates from first separation after the separation of each group you use, because these filtrates may contain cations from other groups. You can refer to the confirmatory tests for each individual cation to be sure of the results you obtained.

II.6.1 Separation of group I

The cations of group I were separated in the form of white precipitate of silver chloride, mercury(I) chloride and lead chloride by adding HCl solution. The concentration should be 0.3N to prevent the precipitation of other cations from other groups, such as Bismuth(III) ions according to the reaction;

$$Bi^{3+} + Cl^- + H_2O \rightarrow BiOCl\downarrow + 2H^+$$

However, if the concentration of HCl is above 0.3N, the cations of silver and lead will not remain precipitated as chloride, however, it will be converted into solutions due to the formation of complexes instead, according to the following reactions;

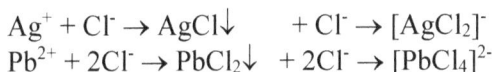

$$Ag^+ + Cl^- \rightarrow AgCl\downarrow \quad + Cl^- \rightarrow [AgCl_2]^-$$
$$Pb^{2+} + 2Cl^- \rightarrow PbCl_2\downarrow \quad + 2Cl^- \rightarrow [PbCl_4]^{2-}$$

We can summarize the procedures of separation and identification in a way to be easy as follows:

By adding HCl, If there is a precipitate (1), that is the evidence of the presence of group I cations. Separate the precipitate (ppt) using the available tools for separation, then wash the ppt by using distilled water and keep the filtrate (1), which may contain cations from other groups

Precipitate (1)	Filtrate (1)
Add distilled water to the precipitate (ppt) and boil. Then filter the hot solution of the mixture.	May contains cations from other groups

| Filtrate (a) If it contains Pb^{2+} ion, three steps | Precipitate (1a) Add 1.0ml of NH$_4$OH to the ppt, | |

must be achieved as evidence of this conclusion:	then separate the ppt from the filtrate.	
1- If there is a white ppt formed by the standing on cooling. 2- The addition of K_2CrO_4 to the filtrate and the formation of yellow ppt. 3- The addition of KI and the formation of yellow ppt. This is a good evidence for the existence of Pb^{2+}.	**Filtrate (1b)** If it contains Ag^+, yellow ppt formed by the addition of KI	**Precipitate (1c)** If this ppt turned black and it is not soluble in diluted HNO_3 this is an evidence for the presence of Hg_2^{2+}

II.6.2 Separation of group II

There are two sets of group II as divided into subgroups, the first group contains (Hg^{2+}, Cu^{2+}, Cd^{2+} and Bi^{3+}; subgroup-1) , the second group contains (As^{3+}, Sb^{3+} and Sn^{4+}; subgroup-2). Concerning our studies, we will concentrate on the first subgroup, and some from the other subgroup, by considering the toxicity of some of the cations of the other subgroup, therefore salts of some cations will be excluded from our studies. The cations of group II were separated in the form of sulphide precipitates (HgS, CuS, CdS, Bi_2S_3, Sb_2S_3, SnS_2) in acidic medium (dilute HCl) to avoid the precipitation of other sulphides, such as in group IV as will as magnesium sulphide. The sulphides of group II are different in their behavior as acid-base character. The precipitates; (HgS, CuS, CdS and Bi_2S_3) are insoluble in alkaline solutions (i.e. KOH), whereas (Sb_2S_3 and SnS_2) can be dissolved in basic medium. These advantages must be established to separate and to differentiate between these subgroups.

By using either the filtrate (1) from group I or new group II mixture solution, add HCl first, then hydrogen sulphide can be added. There are several sources of H_2S. Kip's instrument used (FeS/CaS + Acid $\rightarrow$ H_2S + metal salt). Ammonium sulphide, thioacetamide and other chemicals can be used also. Heat the mixture for the completion of the reaction. The formation of precipitate confirms the evidence the presence of group II cations. Separate the precipitate (2) using the available tools for separation, then wash the ppt by distilled water then keep the filtrate (2) which may contain cations from other groups. Try to dissolve the ppt in KOH solution and separate the precipitate (2) using the available tools for separation, then wash the ppt by excess of KOH solution then keep the filtrate (2a) which may contain the other subgroup II cations.

Precipitate (2)	Filtrate (2a)
Add appropriate amount of dil. HNO_3 to the ppt and boil, then filter the hot solution of the mixture.	May contain subgroup II cations

Precipitate (2a)	Filtrate (2b)		
May contains Hg^{2+}.	May contains Cu^{2+}, Cd^{2+} and Bi^{3+}		
If the ppt is black color and soluble in aqua regia and gave a gray ppt, that confirm the presence of Hg^{2+}	Add to this solution few drops of conc. H_2SO_4, and then heat the mixture for few minutes, until the white fumes appear. Cool and filter the solution [If there is any ppt, that is the evidence of presence of Pb^{2+} which does not completely precipitated during the precipitation of group I cations]. Add excess of NH_4OH then separate the ppt from the filtrate.		
	Precipitate (2b)	Filtrate (2c) (divide into 2 parts)	
	Dissolve this ppt in dil. HCl, and then divide the solution into 2 parts. 1- Add tap water, white ppt will form. 2- Add $NaAsO_2$, a dark brown ppt formed. Both tests confirm the presence of Bi^{3+}	1- If the filtrate is blue in color, add to this part $K_4Fe(CN)_6$. If a brown ppt formed, the two tests confirm the presence of Cu^{2+}.	2- Add KCN solution until the blue color disappeared, and then add H_2S. If a yellow ppt formed, that confirms the presence of Cd^{2+}.

Filtrate (2a)
Add few drops of conc. HCl to this filtrate, then divide the solution into 3 parts

Part 1- Put a piece of iron or silver in the solution then note the formation of black layer. **Part 2**- Fill the tube with tap water and note the formation of white ppt. Both tests Confirm the presence of Sb^{3+}.	Part 3- Add $HgCl_2$ solution, then note the formation of white ppt in presence of magnesium film. This confirm the presence of Sn^{4+}.

II.6.3 Separation of group III

Concerning our studies, after the separation of groups I and II. There are four cations in this group (Fe^{2+}/ Fe^{3+}, Cr^{3+} and Al^{3+}). These cations can be precipitated as hydroxides, which have different colors. By using either a new group III mixture solution or the filtrate (2) which should be treated by adding HCl and boil to remove all H_2S in the solution, then add excess of NH_4Cl solution followed by the addition of NH_4OH solution. The formation of the precipitate confirms the evidence of the presence of group III cations. Separate the precipitate (3) using the available tools, then wash the ppt by distilled water and keep the filtrate (3) which may contain cations from other groups.

Precipitate (3)			Filtrate (3)
Wash the ppt several times with NH_4Cl solution, then transfer the ppt to a beaker and add NaOH solution. Then boil and filter the mixture.			May contain cations from other groups
Filtrate (3a)	**Precipitate (3b)**		
Add solid NH_4Cl then boil the mixture. The formation of gelatinous white ppt that confirms the presence of Al^{3+}.	Transfer the ppt which may contain Fe(II/III) and Cr(III) hydroxides into a beaker, then add about 2g of Na_2O. Diluted the mixture by water. Boil the mixture and filter the contents.		
	Precipitate (3c)	**Filtrate (3d)**	
	Dissolve the ppt in hot conc. HNO_3 (to convert Fe^{2+} ion, if it is present into Fe^{3+}). Diluted the mixture then $K_4Fe(CN)_6$ can be added. If blue ppt	Add CH_3CO_2H to make the solution acidic, and then add $(CH_3CO_2)_2Pb$ solution. A yellow ppt	

	is formed, confirm the presence of Fe^{3+} ion.	confirms the presence of Cr^{3+}.

II.6.4 Separation of Group IV

By using either the filtrate (3) left from group III or new group IV mixture solution. Add NH_4Cl solution then NH_4OH solution. Be sure that there is no cloudiness or ppt formed. Add H_2S (any source), then heat the mixture for the completion of the reaction. The formation of precipitate (4) confirms the presence of group IV cations. This group contains (Mn^{2+}, Zn^{2+}, Co^{2+} and Ni^{2+}) which precipitated as sulphide in basic medium. Separate the precipitate (4) using the available tools, then wash the ppt by distilled water and keep the filtrate (4) which may contain cations from other groups.

Precipitate (4)		Filtrate (4)	
Wash the ppt with NH_4Cl solution, transfer to the beaker then dissolve in hot dilute HCl, Filter the mixture.		May contain cations from other groups	
Filtrate (4a) May contain Zn^{2+} and Mn^{2+}		**Precipitate (4a)** May contain Co^{2+} and Ni^{2+}	
Remove the remaining H_2S gas by boiling, and then add excess of NaOH solution.		Dissolve the ppt in aqua regia, then divide the solution into 2 parts	
Filtrate (4b) Add H_2S or $K_4[Fe(CN)_6]$. A white ppt formed. This confirms the presence of Zn^{2+}.	**Precipitate (4b)** This ppt will start to change to brown color. Add conc. HNO_3 and $NaBiO_2$ or PbO. Purple color confirms the presence of Mn^{2+}.	1- Add NH_4Cl and NH_4OH solutions to make it basic, and then add dimethylglyoxime. The formation of red color confirms the presence of Ni^{2+}.	2- Add CH_3CO_2H Then KNO_2 solution. The formation of yellow ppt confirms the presence of Co^{2+}.

48   *General Practical Chemistry*

II.6.5 Separation of group V

By using either the filtrate (4) left from group IV or new group V mixture solution. Add NH_4Cl then NH_4OH solutions. Be sure that there is no cloudiness or ppt formed. Add $(NH_4)_2CO_3$ solution. The formation of white precipitate confirms the presence of group V cations. This group contains (Ca^{2+}, Ba^{2+} and Sr^{2+}) precipitated as carbonate in basic medium. Separate the precipitate (5) using the available tools, then wash the ppt by distilled water and keep the filtrate (5) which may contain cations from group VI.

Precipitate (5)	Filtrate (5)
Wash the ppt with distilled water then dissolve it in CH_3CO_2H. Add excess of K_2CrO_4 solution to the mixture. Note the formation of yellow ppt. Filter the mixture.	May contain cations from group VI

Precipitate (5a)	Filtrate (5a)	
On cold, the addition of K_2CrO_4 solution leads to the formation of yellow ppt which confirms the presence of Ba^{2+} ion.	Add to this mixture dil. H_2SO_4, C_2H_5OH then boil and Filter.	
	The white ppt confirms the presence of Sr^{2+} ion.	The filtrate contains Ca^{2+} ion, if there is no white ppt from $CaSO_4$ solution

II.6.6 Separation of group VI

By using either the filtrate (5) left from group V or new Group VI mixture solution. Since there is no specific group reagent for this group which contain (K^+, Na^+, NH_4^+ and Mg^{2+} ions). The following steps should be followed. Divide the filtrate (5) if you use this filtrate left from separation of group V which precipitated as carbonate in basic medium. Separate the precipitate (5) using the available tools, then wash the ppt by distilled water and keep the filtrate (5) which may contain cations from group VI.

Part-1	Part-2
Add few drops of NH₄OH solution and then excess of Na₃PO₄solution. Worm the mixture or scratch the wall of the tube or the beaker. If a white ppt is formed that confirm the presence of Mg^{2+} ion.	Some of NH₄OH solution remains from the previous steps of separation should be removed by boiling the mixture in a beaker until dryness. If there is no solid residue left. This confirm the absence of Na^+ and K^+ ions. If the solid residue left, the residue should be dissolved in small amount of distilled water then filter the clear filtrate. 1- To small amount of the filtrate add sodium cobaltinitrite $Na_3[Co(NO_2)_6]$, If a yellow ppt is formed, that confirm the presence of K^+ ion. 2- Flame test should be preformed to detect whether K^+ or Na^+ present or not.

II.7 The most important chemical reactions of basic radicals

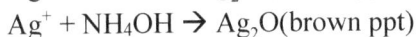

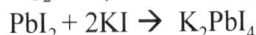

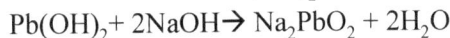

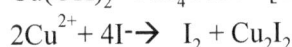

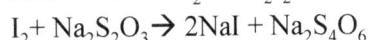

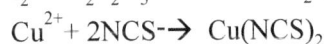

$Ag^+ + NaOH \rightarrow Ag_2O$ (brown ppt)

$Ag^+ + NH_4OH \rightarrow Ag_2O$(brown ppt)

Ag_2O(brown ppt) + excess $NH_4OH \rightarrow 2Ag(NH_3)_2OH$

$AgI + KI \rightarrow KAgI_2$

$Hg_2I_2 + KI \rightarrow K_2[Hg(I)_4] + Hg_{(s)}$

$Hg_2CrO_4 + heat \rightarrow HgCrO_4$

$PbI_2 + 2KI \rightarrow K_2PbI_4$

$Pb^{2+} + 2NaOH \rightarrow Pb(OH)_2$

$Pb(OH)_2 + 2NaOH \rightarrow Na_2PbO_2 + 2H_2O$

$Cu^{2+} + NaOH \rightarrow Cu(OH)_2 \rightarrow CuO$

$Cu^{2+} + NH_4OH \rightarrow Cu(OH)_2$

$Cu(OH)_2 + NH_4OH \rightarrow [Cu(NH_3)_4]^{2+}$

$2Cu^{2+} + 4I- \rightarrow I_2 + Cu_2I_2$

$I_2 + Na_2S_2O_3 \rightarrow 2NaI + Na_2S_4O_6$

$Cu^{2+} + 2NCS- \rightarrow Cu(NCS)_2$

$3Cu^{2+} + 2Al \rightarrow 2Al^{3+} + 3Cu$

$Cd^{2+} + NaOH \rightarrow Cd(OH)_2$

$Cd^{2+} + NH_4OH \rightarrow \mathbf{Cd(OH)_2}$

$Cd(OH)_2 + NH_4OH \rightarrow [Cd(NH_3)_4]^{2+}$

$Hg^{2+} + NaOH \rightarrow Hg(OH)_2$

$Hg(OH)_2 + \text{excess } NaOH \rightarrow HgO$

$Hg^{2+} + NH_4OH \rightarrow Hg(NH_2)Cl$

$Hg^{2+} + I_2 \rightarrow HgI_2$

$HgI_2 + KI \rightarrow K_2HgI_4$

$Hg^{2+} + Sn^{2+} + 2Cl^- \rightarrow \mathbf{Hg_2Cl_2} + Sn^{4+}$

$Bi^{3+} + NaOH \rightarrow Bi(OH)_3$

$Bi(OH)_3 + NaOH \rightarrow BiO(OH) + H_2O$

$Bi^{3+} + NH_4OH \rightarrow Bi(OH)_3$

$Bi^{3+} + 3I^- \rightarrow BiI_3$

$BiI_3 + KI \rightarrow KBiI_4$

$Bi^{3+} + NCS^- + \mathbf{H_2O} \rightarrow Bi(SCN)O$

$Sb^{3+} + NaOH \rightarrow Sb_4O_6$

$2Sb^{3+} + 3Zn \rightarrow 3Zn^{2+} + 2Sb$

$Sn^{4+} + 4NaOH \rightarrow Sn(OH)_4 + 4Na^+$

$Sn(OH)_4 + 2NaOH \rightarrow Na_2SnO_3 + 3H_2O$

$Sb_4O_6 + 4NaOH \rightarrow 4NaSbO_2 + 2H_2O$

$Sn^{4+} + 2Zn \rightarrow 2Zn^{2+} + Sn$

$Sn^{4+} + Fe/HCl \rightarrow \mathbf{Sn^{2+}}$

$Sn^{2+} + \mathbf{2HgCl_2} \rightarrow \mathbf{Sn^{4+}} + \mathbf{Hg_2Cl_2}$

$As^{3+} + AgNO_3 \rightarrow AgAsO_3$

$AgAsO_3 + NH_4OH \rightarrow [Ag(NH_3)_2]^+ + (NH_4)_3AsO_3$

$AS^{3+} + I_2 \rightarrow As^{5+} + 2I^-$

$2As^{3+} + 3Sn^{2+} \rightarrow 2As + 3Sn^{4+}$

$Al^{3+} + 3NaOH \rightarrow Al(OH)_3 + 3Na^+$

$Al(OH)_3 + NaOH \rightarrow NaAlO_2 + 2H_2O$

$Al^{3+} + 3NH_4OH \rightarrow Al(OH)_3 + 3NH_4$

$Al(OH)_3 + NH_4OH \rightarrow (NH_4)AlO_2 + 2H_2O$

$Al^{3+} + 3Na_2CO_3 \rightarrow Al(OH)3$

$2Al(OH)_3+Na_2CO_3 \rightarrow 2NaAlO_2+CO_2+3H_2O$

$Al^{3+} + PO_4^{3-} \rightarrow AlPO_4$

$AlPO_4 + H^+ \rightarrow Al^{3+} + HPO4^{2-}$

$Cr^{3+} + 3NaOH \rightarrow Cr(OH)_3 + 3Na^+$

$Cr(OH)_3 + 3H^+ \rightarrow Cr^{3+} +3H_2O$

$Cr^{3+} + 3NH_4OH \rightarrow Cr(OH)_3 + 3NH_4^+$

$Cr(OH)_3 +NH_4OH \rightarrow (NH_4)CrO_2+2H_2O$

$Cr(OH)_3 + NaOH + heat \leftrightarrows NaCrO_2 + 2H_2O$

$Cr^{3+} + PO4^{3-} \rightarrow CrPO_4$

$CrPO_4 + H^+ \rightarrow Cr^{3+} + HPO_4^{2-}$

$Fe^{3+} + 3NaOH \rightarrow Fe(OH)_3+ 3Na^+$

$Fe(OH)_3 + 3H^+ \rightarrow Fe^{3+} + 3H_2O$

$Fe^{3+} + 3NH_4OH \rightarrow Fe(OH)_3 + 3NH_4^+$

$Fe^{3+} + PO_4^{3-} \rightarrow FePO_4$

$FePO_4 + H^+ \rightarrow Fe^{3+} + HPO_4^{2-}$

$Fe^{3+} + KSCN \rightarrow Fe(SCN)^{2+} + K^+$

$Fe(SCN)^+ + 2KSCN \rightarrow Fe(SCN)_3$

$Zn^{2+} + 2NaOH \rightarrow Zn(OH)_2 +2Na^+$

$Zn(OH)_2 + 2NaOH \rightarrow Na_2ZnO_2+ 2H_2O$

$Zn^{2+} + 2NH_4OH \rightarrow Zn(OH)_2 + 2NH_4^+$

$Zn(OH)_2 + 4NH_4OH \rightarrow [Zn(NH3)_4]^{2+}$

$Mn^{2+} + 2NH_4OH \rightarrow Mn(OH)_2 + 2NH_4^+$

$2MnSO_4 + 5PbO_2 + 6HNO_3 \rightarrow 2HMnO_4 + 2PbSO_4 + 3Pb(NO_3)_2 + 2H_2O$

$2MnSO_4+ 5NaBiO_2 + 16HNO_3 \rightarrow 2HMnO_4 + 2Na_2SO_4 + 5Bi(NO_3)_3 + NaNO_3 + 7H_2O$

$Ni^{2+} + 2NaOH \rightarrow Ni(OH)_2+ 2Na^+$

$Ni^{2+} + 2NH_4OH \rightarrow Ni(OH)_2 + 2NH_4^+$

$Ni(OH)_2 + 6NH_4OH \rightarrow [Ni(NH_3)_6]^{2+}$

$Ni^{2+} + 2HDMG \rightarrow Ni(DMG)_2 + 2H^+$

$Co^{2+} + 2NH_4OH \rightarrow Co(OH)_2 + 2NH_4^+$

$Co(OH)_2 + 6NH_4OH \rightarrow [Co(NH_3)_6]^{2+}$

$6 NO_2^- + Co^{3+} \rightarrow [Co(NO_2)_6]^{3-}$

$Co^{2+} + NO_2^- + 2H^+ \rightarrow Co^{3+} + NO + H_2O$

$M^{2+} + K_2CrO_4 \rightarrow MCrO_4 + 2 K^+$

$M^{2+} + CaSO_4 \rightarrow MSO_4 + Ca^{2+}$

$M^{2+} + (NH_4)_2C_2O_4 \rightarrow MC_2O_4 + 2NH_4^+$

$NH_4^+ + NaOH \rightarrow NH_3 + H_2O + Na^+$

$NH_3 + HCl \rightarrow NH_4Cl$

$NH_4^+ + K_2Hg_2I_3 + H_2O \rightarrow NH_2Hg_2I_3 + 2KI + H^+ + HI$

$NH_4^+ + Na_3[Co(NO_2)_6] \rightarrow (NH_4)_3[Co(NO_2)_6] + H^+$

$Mg^{2+} + NH_4Cl + Na_2HPO_4 \rightarrow Mg(NH_4)PO_4 + H^+ + Cl^- + 2Na^+$

$Mg^{2+} + CO_3^{2-} \rightarrow MgCO_3$

$3K^+ + Na_3[Co(NO_2)_6] \rightarrow K_3[Co(NO_2)_6] + 3Na^+$

$K^+ + ClO_4^- \rightarrow KClO_4$

$Na^+ + [3UO_2(CH_3CO_2)_2, \quad Mg(CH_3CO_2)_2, \quad CH_3CO_2H] \quad \rightarrow$
$NaMg(UO_2)_3(CH_3CO_2)_6$

Questions

(Acid radicals)

1. Give the correct oxidation state for the following anions:
B_4O_7, NO_2, NO_3, PO_4, $C_2H_3O_2$, CO_3, S_2O_3, Br, Cl and SO_4.
2. Write the names of each of the following compounds:
Cu_2I_2, PbS, Na_2S, $FeCl_3$, Na_3PO_4, $Na_2B_4O_7$, $Ag_2S_2O_3$, $Hg_2(NO_3)_2$, $Ca_3(PO_4)_2$.
3. Fill in the blanks:
- Glycerol test is used for detection of ....
- Bromide reacts with concentrated sulphuric acid and give ....
- Sulphide reacts with lead acetate and form ....
- Carbonate gives white ppt. with ....
- Silver nitrate solution gives canary yellow ppt. with ....
- Bicarbonate gives reddish brown ppt. with ....after heating.
- On addition of nitrite solution to acidified permanganate solution, the color of the permanganate will be.......
- On the addition of I_2 solution to thiosulphate solution, the color of the mixture will be   ....
- The acid radical which gives violet vapor when react with conc. H_2SO_4 is ....
- The acid radical which gives brown gas and pale blue solution when react with dilute HCl is ....
4. By balanced chemical equation, how can you distinguish between the following terms?
- Sulphate and borate
- Carbonate and bicarbonate
- Bromide and iodide
- Nitrite and thiosulphate
- Phosphate and sulphate
5. Discuss the following reactions:
(i) Reaction of potassium carbonate with mercuric chloride.
(ii) Reaction of barium borate with mineral acids.
(iii) Reaction of sodium thiosulphate with acidified potassium permanganate solution.
(iv) Reaction of nitrite with dil. HCl.
(v) Reaction of sodium chloride with conc.H_2SO_4 and ammonia.
(iv) Reaction of lead (II) bromide with hot water.
6. Give the correct color for each of the following compounds:
$MgCO_3$, Cu_2I_2, $MgSO_4$, Ag_3PO_4, PbI_2, $Ba(NO_3)_2$, $FeCl_3$, $KMnO_4$, Conc. H_2SO_4 and H_2O.

7. Write the chemical formula for each of the following compounds: Potassium dichromate, sodiumdihydrogenphosphate, potassium sulphite, bismuth sulphide, hydrogenperoxide, dimethylglyoxime, sodium borate, sodium hypochlorate, and Potassium aluminum sulphate.

8. Complete the following table:

Name	Formula	Formula	Name
Phosphate		PbS_2O_3	
Nitrite		KBr	
Iodide		$K_3Fe(CN)_6$	

9. Complete and give the color of the products in the following chemical reactions:

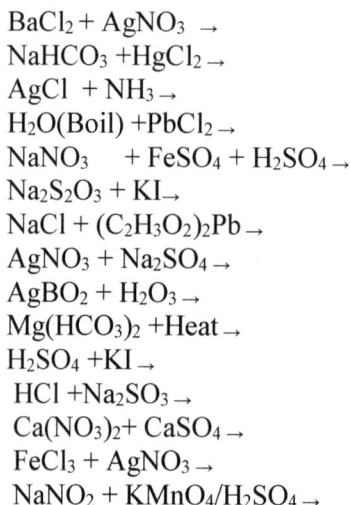

$BaCl_2 + AgNO_3 \rightarrow$

$NaHCO_3 + HgCl_2 \rightarrow$

$AgCl + NH_3 \rightarrow$

$H_2O(Boil) + PbCl_2 \rightarrow$

$NaNO_3 + FeSO_4 + H_2SO_4 \rightarrow$

$Na_2S_2O_3 + KI \rightarrow$

$NaCl + (C_2H_3O_2)_2Pb \rightarrow$

$AgNO_3 + Na_2SO_4 \rightarrow$

$AgBO_2 + H_2O_3 \rightarrow$

$Mg(HCO_3)_2 + Heat \rightarrow$

$H_2SO_4 + KI \rightarrow$

$HCl + Na_2SO_3 \rightarrow$

$Ca(NO_3)_2 + CaSO_4 \rightarrow$

$FeCl_3 + AgNO_3 \rightarrow$

$NaNO_2 + KMnO_4/H_2SO_4 \rightarrow$

(Basic radicals)

1. What is the valence of the cations of the elements; Ba, Cu, Cd, Ni, Ca, Mg, NH_4, Pb, Cr and Bi.

2. Write the chemical formula for the following compounds: Sodium aluminate, Potassium thiocyanate, Potassium ferricyanide, Sodium hydroxide, Sulphuric acid, Bismuth thiosulphate, potassium ferrocyanide, Calcium hydroxide, Disodium hydrogen phosphate and mercurous nitrate.

3. Complete the following blanks:

- $NH_4Cl + NH_4OH + (NH_4)_2CO_3$ are a group reagent for group......, which contains......, ...,.....ions.

- The colors of $Cr(OH)_3$ and $Al(OH)_3$ compounds are ...., ..., while $PbCl_2$ has a ...... ppt.
- The flame test for Ba^{2+} has a ....color as for Na^+ is ....
- Dilute $HCl + H_2S$ is a group reagent for group ......which contains ......, ...,...... ions.
- The color for CuS and Bi_2S_3 compounds are ...., while the ppt for Sb^{3+} is ......
- The roles of NH_4Cl and NH_4OH in the group three group reagent are ....
- The flame test for K^+ ion is ...... and ..... for Ba^{2+} ion and .... for NH_4^+ ion.
4. By special chemical reaction, how to distinguish between
(a) Mn^{2+} and Zn^{2+} ions. (b) Co^{2+} and Ni^{2+} ions
5. Define the group reagent for each of the following cations:
Sr^{2+}, Fe^{3+}, Cu^{2+}, Cd^{2+}, Mn^{2+}, Ca^{2+}, Mg^{2+}, NH_4^+, Pb^{2+}, Cr^{3+}.
6. Discuss the confirmation tests of cations of group 6.
7. By using K_2CrO_4 solution, how you distinguish between cations of group one.
8. What is the main difference between cations of group two and four?
9. What happen if you add ammonia solution to the silver chloride ppt?
10. Complete and balance the following chemical reactions:

$KI + Hg_2(NO_3)_2 \rightarrow$

$Pb(NO_3)_2 + H_2SO_4 \rightarrow$

$Fe(NO_3)_3 + K_4[Fe(CN)_6] \rightarrow$

$NH_4NO_3 + NaOH \rightarrow$

$H_3O^+ + Cl^- \rightarrow$

$Ba(NO_3)_2 + CaSO_4 \rightarrow$

$Cu(NO_3)_2 + H_2S \rightarrow$

$Hg_2(NO_3)_2 + K_2CrO_4 \rightarrow$

Chapter III

Qualitative Organic Chemical Analysis

III.1 Techniques

1. Melting and boiling point determination

The purpose of this experiment is to determine the melting points of a series of compounds and then to carry out the partial identification of an organic compound by means of mixed melting point test. Arrange the Thiele tube melting point apparatus. Make a small lengthwise slit in the stopper used for supporting the thermometer so that the stem is exposed throughout. Add cottonseed oil (or paraffin) to a level just above the top of the side arm. If necessary, powder the sample by means of a mortar and pestle or by crushing on a small filter paper with a spatula. Some organic compounds of known melting and boiling points are listed in table 3.1.

As a liquid is heated, the vapor pressure of the liquid increases to the point where it just equals the applied pressure (usually atmospheric pressure). At this point, the liquid will be observed to boil. The normal boiling point is measured at 760 mmHg (1.0 atmosphere) At a lower applied pressure, the vapor pressure needed for boiling is also lowered, and the liquid will now boil at a lower temperature. The relationship between the applied pressure and the temperature of boiling for a liquid is determined by its vapor-pressure-temperature behavior of a liquid. The experiments will be considered in the distillation techniques.

2. Sublimation

Sublimation is usual a property of relatively nonpolar substances that also have highly symmetrical structure which has relatively high melting point and high vapor pressure. The ease with which a substance can escape from the solid is determined by the strength of intermolecular forces (symmetrical molecular density and small dipole moment). A smaller dipole moment means a higher vapor pressure because of lower electrostatic forces in the crystal. Solids sublime if their vapor pressure are sizable at pressure at solid at m.p. = 780 mmHg. Camphor m.p. = $179°$ and vapor pressure at m.p. = 370 mmHg,

Naphthalene m.p. = 80° and vapor pressure at m.p. = 7 mmHg and
Benzoic acid m.p. =122° and vapor pressure at m.p. = 6 mmHg.

Table 3.1 Melting and boiling points of common compounds

	Compound	M.P.	B.P.		Compound	M.P.	B.P.
1-	Acetaldehyde	-	21	33-	Glucose	83	-
2-	acetamide	81	-	34-	Heptane	-	89.4
3-	Acetanilide	115	-	35-	Hexane	-	69
4-	Acetic acid	16.6	118	36-	Hexanedioic acid	152	-
5-	Acetic anhydride	-	139	37-	Methanol	-	65
6-	Acetone	-	56.5	38-	2-Methoxyphenol	32	204
7-	Adipic acid	152		39-	N-Methylaniline	-	196
8-	Aniline	-	184	40-	Methylformate	-	31.5
9-	Anthranilic acid	146		41-	3-Methylphenol	-	191
10-	Benzaldehyde	-	179	42-	4-Methylphenol	-	202
11-	Benzamide	129	-	43-	Naphthalene	80.2	-
12-	Benzanilide	161	-	44-	2-Nitrophenol	45	216
13-	Benzoic acid	122	250	45-	3-Nitrosalicylic a.	144	-
14-	Benzoin	137	-	46-	Oxalic acid	101	-
15-	Benzophenone	48	305	47-	3-Pentanone	-	101
16-	m-Bromobenzoic acid	156	-	48-	Petroleum ether	-	40-60
17-	1-Butanol	-	117.5	49-	Phenacetin	135	-
18-	Butylamine	-	78	50-	Phenol	-	181
19-	o-Chlorobenzoic acid	140	-	51-	Phenyl acetate	-	196
20-	Cinnamic acid	133	300	52-	Phenylacetone	27	216
21-	Cyclohexane	-	80.7	53-	Phthalamide	221	-
22-	Cyclohexanol	-	161	54-	Phthalic anhydride	131	-
23-	Cyclohexanone	-	156	55-	Salicylic acid	158	-
24-	Cyclohexene	-	83	56-	Styrene	-	156
25-	Diethylaniline	-	215.5	57-	Succinic acid	188	-
26-	N,N-Dimethyl aniline	-	193	58-	Thiourea	178	-

27-	2-Ethoxyethyl acetate	-	156		59-	Toluene	-	110.6
28-	Ethyl acetate	-	77		60-	Trans-Cinnamic a.	133	-
29-	Ethyl alcohol	-	78		61-	Trichloromethane	-	76.7
30-	Ethyl benzoate	-	212		62-	Urea	132.7	-
31-	Ethylenediamine	-	117		63-	p-Toluidine	251	
32-	Ethyl methyl ketone	-	80					

Furthermore, vapor pressure increased with a decreased atmospheric pressure. Therefore, some compounds sublimated under reduced pressure.

Sublimation is a process occasionally used for purification of solid organic compounds. Its use is necessarily limited to those compounds which on heating pass readily and directly from the solid to the vapor state, with a subsequent ready reversal of this process when the vapor is cooled.

Procedures

A simple form of apparatus which gives good results consist of a small evaporating basin in which 10 g of benzoic acid is placed and covered with a filter paper. The filter paper is placed by a number of small holes (about 2 mm in diameter) made as powder direction, a glass funnel with smaller diameter than the diameter at the basin is put inverted over the paper. The basin is then gently heated on wire gauze by a small Bunsen flame, which should be carefully protected from side draughts so that the material in the basin receives a steady uniform supply of heat. The benzoic acid vaporizes and the vapors passes through the holes into the cold funnel and condense as fine crystals on the upper surface of the paper and on the walls at the funnel .When almost the whole at the benzoic acid has vaporized stop the heating and the pure sublimed benzoic acid is collected. Determine m.p. of the crude and pure benzoic acid.

3. Distillation

The separation of organic liquid compounds is one of the most important tasks of the organic chemist. Organic compounds seldom occur in pure form in nature or as products of a laboratory preparation (or synthesis). The most commonly used method for purification of liquids is distillation, a process by which one liquid can be separated from another liquid or a liquid from a nonvolatile solid. When water is heated in a simple distillation apparatus, the vapor pressure of the liquid or the tendency of molecules to escape from the surface, increases, until it becomes equal to the atmospheric pressure, at which the liquid begins to boil. Addition of more heat will supply the heat of vaporization required for conversion of liquid water to gas (steam), which rises in the apparatus, warms the distillation head and thermometer, and flow down the condenser. The cool walls of the condenser remove heat from the vapor and vapor condenses to the liquid from. Distillation should be conducted slowly and steadily and at rate such that the thermometer bulb always carries a drop of condensate and is bathed in a flow of vapor. Liquid and vapor are then in equilibrium, and the temperature recorded is the true boiling point. If excessive heat is applied to the walls of the distilling flask above the liquid level, the vapor can become superheated, the drop will disappear from thermometer, the liquid-vapor equilibrium is upset and the temperature of vapor rises above b.p. It is never possible to separate a mixture completely by a simple distillation. However, in two cases it is possible to get an acceptable separation into relatively pure components. In the first case, if the boiling points of A and B differ by a large amount (>100) it will be possible to get a fair separation of A and B. In the second case, if A contains small amount of B (<10%). When boiling -point differences are not large, and when highly pure compounds are desired, it is necessary to do a fractional distillation.

Procedures

In a 250 ml distillation flask, a 100ml sodium chloride solution is placed with few fragments of porcelain-chip. Then the distillation flask is capped with distilling head which is fitted to a water-condenser and a thermometer (250 °C). The other end of the condenser is fitted to an adapter which then fitted to conical flask 250 ml. Heat the mixture on wire gauze by Bunsen flame. When the boiling point of the first liquid to be distilled is reached a ring of condensate (or reflux ring) move up through the apparatus and meets the thermometer bulb. At this point, the thermometer records rapped rise in temperature and soon condenser begins to pass from the condenser into the receiver collecting the first fraction while the temperature maintained being constant. Write the volume and b.p. of water.

4. Decolourization

The crude products of an organic reaction may contain colored impurities. These impurities can be removed by boiling the substance in solution with a little decolorizing charcoal for 5-10 minutes and then filter the solution while it is hot.

5. Recrystallization

Solid organic compounds when isolated from organic reactions are seldom pure. The purification of impure crystalline compounds is usually affected by crystallization from a suitable solvent or mixture of solvents. The purification of solids by crystallization is based upon differences in their solubility in a given solvent or mixture of solvents. In its simplest form the crystallization process consists of:

1) Dissolve the impure substance in suitable solvent at or near the boiling point.
2) Filter the hot solution from particles of insoluble material and dust.
3) Allow the hot solution to cool thus causing the dissolved substance comes to crystallize.
4) Separate the crystals from mother liquor, then the solid dried and tested for purity by m.p. determination.

The most desirable characteristics of solvent for recrystallization are as follows:

1) A high solvent power for the substance to be purified at elevated temperatures and a comparatively low solvent power at room temperature or below.
2) It should dissolve the impurities readily or to only a very small extent.
3) It should yield well - formed crystals of the purified compound.
4) It. must be capable of easy removal from the crystals of the purified compound i.e. posses a relatively low boiling point.

The following rough generalizations may assist in the selection of the recrystallization solvent.

1) Substance is likely to be most soluble in a solvent to which it is most closely related in chemical and physical characteristics.

2) A polar substance is more soluble in polar solvents and less soluble in non-polar solvents. In ascending a homologous series, the solubility of the members tends to become more and more like that of the hydrocarbon from which they may be regarded as being derived.

- Choice of solvents for recrystallization

Experiment (Bring 3 samples: salicylic acid, acetanilide and naphthalene)
Place 0.1 g of one of the above substances in a test tube and add the solvent (C_2H_5OH or H_2O) a drop at time with continues shaking. After 1ml of the solvent has been added, the mixture is heated to boiling. Precautions should be taken if the solvent is flammable. If the sample dissolves easily in 1ml of cold solvent or upon gentle warming, the solvent is unsuitable. If the solid is not dissolved, more solvent is added in 0.5ml portions and again heated to boiling after each addition. If 3ml of solvent is added and the substance does not dissolve on heating, the substance is regarded as sparingly soluble in that solvent another solvent should be sought. If the compound dissolves in hot solvent and on cooling crystallization is occurred, so this is the proper solvent. Finally summarize your results and indicate the most suitable solvent or solvents for the recrystallization of each of the chosen compounds.

- Recrystallization of acetanilide

Weigh out 4.0 g of acetanilide into a 250 ml conical flask. Add 80ml of water and heat nearly to the boiling point. The acetanilide will appear to melt as oil form in the solution. Add small portions of hot water, while stirring the mixture and boiling until the solid is dissolved gently. If the solution is not colorless, allow to cool slightly, add 0.1 g of discoloring carbon and continue the boiling for a few minutes in order to remove the colored impurities. Filter the boiling solution through a fluted filter paper supported in a short-necked funnel and collect the filtrate in a 250 ml conical flask. Cover the flask containing the hot filtrate with a clock glass and cool. Allow to stand for about 30min. to complete the separation of the solid. Filter with suction through a small Buchner funnel; wash the crystals twice with 5 ml cold water. Allow the crystals to dry in the air. For more rapid drying you can follow other procedures, the crystals may be placed in an oven held at temperature of about 80 °C. Calculate the yield. Determine the m.p. before and soon after recrystallization.

6. Extraction

No technique is more widely used for the separation of an organic product from its reaction mixture or for the isolation of naturally occurring organic substances than extraction. We may define the extraction as the separation of a component from a mixture by means of a solvent. In practice, extraction is usually employed to separate an organic compound from an aqueous solution or suspension. The process consists of shaking the water solution or suspension with a water-immiscible organic solvent and allowing the layers to separate. The various solutes present distribute themselves between the aqueous and organic layers according to their relative solubility.

General Rules: If we have a mixture of (R-C_6H_4-OH, RCOOH, RNH_2, RCOR), the general scheme will be as follows (using aqueous solution for HCl, NaOH and $NaHCO_3$ as reagents):

```
                                                              ether    RCOR
 RC6H4OH            water   RCOONa                            layer
 RCOOH     Ether    layer                  ether   RNH2  ether
 RNH2      NaHCO3                           layer   RCOR  HCl
 RCOR              ether   RC6H4OH  ether                     water    RNH3Cl
                   layer   RNH2     NaOH                      layer
                           RCOR             water   RC6H4ONa
                                            layer
```

III.2 Identification and reactions

III.2.1 Physical properties

1- Condition: Organic compounds may be a gas, liquid or solid.

Solids: They are classified as crystalline (plates, prisms, needles, microcrystalline powder) or amorphous (no definite shape). Many organic compounds can have two crystallized forms (dimorphism or present in more than two forms (polymorphism) the shape of crystals may give evidence about the structure of the molecule (e.g. Tartaric acid; **D** and **L** forms are enantiomorphism, i.e. crystals are mirror image to each other).

Liquids: compounds may be mobile such as methyl alcohol, ethyl alcohol) or viscous such as glycerol.

2- Color: There is relation between the color and the structure of the organic compounds. Certain groups produce a color such as $-NO_2$, $-NO$, $-N_2$ groups or extended conjugation (chromophores). Other groups when present together with chromophoric groups, the color increases (e.g. OH and NH_2) (auxochroms). The original color should be differentiated from the coloration due to impurities or partial oxidation as in case of $C_6H_5NH_2$, is colorless in pure state but usually brown in color due to auto-oxidation.

3- Odor: The odor of organic compounds may indicate the presence of certain groups of compounds (e.g.: Esters and ethers are usually had pleasant or fruity in odor, anhydrides, side chain halogenated compounds are possessing pungent odor. Benzaldehyde, nitrobenzene and cyanobenzene possess bitter almonds odor. Monohydric phenols and nitrophenols, have phenolic odor. Methyl and ethyl esters of salicylic acid possess the odor of green oil of winter. p-toluidine, Stable-like odor and acetamide, indicating by Mice-like odor.

III.2.2 Solubility

1- Solubility in H₂O: Many organic compounds, which have polar functional groups, and few numbers of carbon atoms (say$<C_6$-C_1) will be soluble or partially soluble in H_2O. Many carboxylic acids and most salts of carboxylic acids are soluble in H_2O.

Procedures: Place 0.2 ml (0.1 g of a solid) of the compound in a small test tube and add in portions 3ml of water. Shake vigorously after the addition of each portion of solvent, being careful to keep the mixture at room temperature. If the compound dissolves completely, record it as soluble. Solids should be finely powdered to increase the rate of solubility. If the solid appears to be insoluble in water or ether, it is sometimes advisable to heat the mixture gently. Acid-base properties of water-soluble compounds should be determined with litmus paper.

2- Solubility due to the reaction: The most organic compounds, which are insoluble in cold water, show their chemical nature due to the reaction with solvents. The following scheme and table 3-2 summarize the solvent and the organic compounds classifications according to their behavior in the solubility. Common solvents used for the solubility: 5-10% (HCl, NaOH, $NaHCO_3$), conc. H_2SO_4 beside organic solvents.

Table 3.2 Solubility groups of organic compounds

I/H₂O+ether	II/H₂O not ether	III A/ 5%NaOH+5%NaHCO₃	III B/ 5% NaOH not 5%NaHCO₃	IV/ 5% HCl	V/ Conc. H₂SO₄	VI/ no group reagent
Alcohols, amines, aldehydes, ketones, carboxy acids, phenols	Amides, sugars, polycarboxy acids	Caboxy acids	Phenols	Amines	Aldehydes, ketones, amides, esters, unsaturated hydrocarbones	Haloalkanes, saturated hydrocarbones

III.2.3 Action of dry heat

Heat a small portion or few ml in crucible or piece of porcelain and note the following:

1- The change of the appearance
2- The degree of inflammability
3- The odor of product
4- The remaining not-volatile residue

a) Non-luminous flame indicates low percentage of carbon as aliphatic compounds.
b) Luminous smoky flame indicates high %carbon a highly unsaturated either aliphatic or aromatic compounds.
c) The aromatic nature of the compounds is confirmed by nitration (heat 0.1g or few ml of the organic compound in dry test tube with few ml of a mixture of equal volume of conc. H_2SO_4 and HNO_3. Then the mixture is poured in ice cooled water and then a yellow oily or solid nitro compound is separated. Nitro phenols and aromatic amines will be unchanged by the nitration, so this test is of no significance. Some compounds are non-inflammable, such as polyhalogenated derivatives (e.g.: chloroform, carbon tetrachloride and those compounds containing metal).

Residual ashes: The residual ash could be carbon and it is dissolved by few drops of conc. HNO_3 and heat.

Metallic residual ash does not dissolve in conc. HNO_3 and heat. The metallic residue is either due to salts of carboxylic acids or sulphonic acid.

Carbonization or charring of organic compounds: This occurs in carbohydrates or aliphatic compounds.

Hydroxyl acids (e.g.; citric or tartaric acid): they give burnt sugar odor.

Alcohols: Burns rapidly with a clear flame, typical of many aliphatic substances.

Urea: Melts then boil then solidify (if a solution of the resulting solid is treated with Beurit's test, a purple color will form. This is differentiating urea as amide from acetamide and benzamide.

Aniline: A very smoky flame, typical of many aromatic compounds.

Chloroform: Does not burn, the vapor becomes hot and then burns with a slightly smoky flame, typical of substance rich in halogens.

Salts (e.g. Na benzoate) Burns with difficulty leaving residue. When heated with conc. HNO_3 carbonaceous matter is oxidized leaving a white infusible residue, typical of alkali metals salts of carboxylic acids

Sugars: Melts, darken, then chars and finally burn with a marked odor of burnt sugar.

Tartaric acid and tartrates: Swell up, blacken and then give burnt sugar odor.

Citrates and lactates: Char and then give burnt sugar odor.

Starch: Leaves a hard-black residue of carbon.

III.2.4 Action of soda lime (NaOH/CaO) for solids only:

The action of this reagent is to decompose or dehydrate the substances (mix 1 part of the substance with 3 parts of soda lime. Triturate in a mortar (without heating).
   - Ammonium salts give ammonia odor on cold
   - Chloral hydrate gives the odor of $CHCl_3$ odor
   - Aniline salt gives the aniline odor
On heating (place the mixture in a test tube and add another layer of soda lime (the upper layer of soda lime acts as a safeguard to grad

against escapement of any unreacted substance heated). Soda lime when applied as such acts as:

1- Hydroxylizing agent

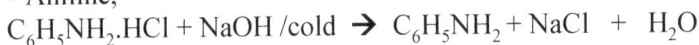

- Ammonium salts: Will produce NH_3 which can turn moister litmus paper from red to blue.
$RCOONH_4$+ soda lime /cold ➔ $RCOONa$+ NH_4OH
- Amides and imides on heating give ammonia:
$RCONH_2 + NaOH$ /Δ ➔ $RCOONa + NH_3$
- Aniline,
$C_6H_5NH_2.HCl + NaOH$ /cold ➔ $C_6H_5NH_2 + NaCl + H_2O$
- Phthalimide and bezaminde: Bitter almond odor beside the ammonia odor, due to formation of bezonitrile

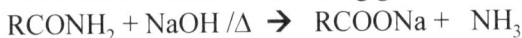

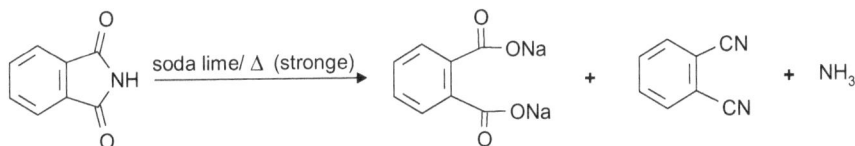

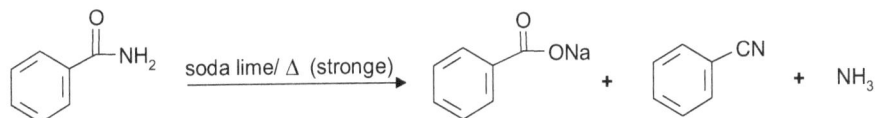

2- Decarboxylating agent

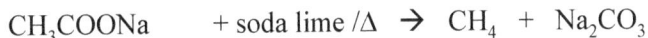

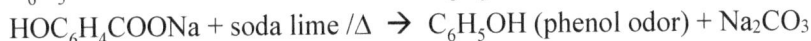

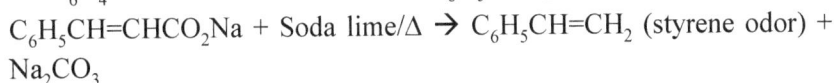

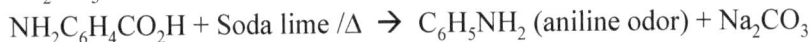

CH_3COONa + soda lime /Δ ➔ $CH_4 + Na_2CO_3$
$(CH_2COONa)_2$ + soda lime /Δ ➔ $CH_3CH_3 + 2Na_2CO_3$
C_6H_5COONa + soda lime /Δ ➔ C_6H_6 (benzene odor) + Na_2CO_3
HOC_6H_4COONa + soda lime /Δ ➔ C_6H_5OH (phenol odor) + Na_2CO_3
$C_6H_5CH=CHCO_2Na$ + Soda lime/Δ ➔ $C_6H_5CH=CH_2$ (styrene odor) + Na_2CO_3
$NH_2C_6H_4CO_2H$ + Soda lime /Δ ➔ $C_6H_5NH_2$ (aniline odor) + Na_2CO_3
antharanilic acid

$NH_2C_6H_4SO_3H$ + Soda lime /Δ → $C_6H_5NH_2$ (aniline odor) + Na_2SO_3
sulphanilic acid

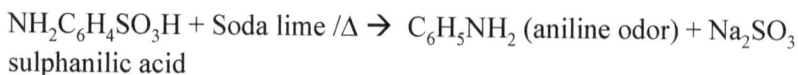

3- Dehydrating agent

Carbohydrates, aliphatic hydroxy acids (tartaric and citric acids or their salts) + Sodalime /Δ → burnt sugar odor

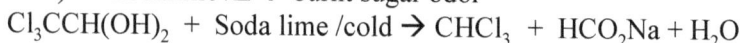

$Cl_3CCH(OH)_2$ + Soda lime /cold → $CHCl_3$ + $HCO_2Na + H_2O$

III.2.5 Action of sodium carbonate solution

Place 5ml of dilute sodium carbonate solution in a test tube, add few fragments of porcelain and boil gentle to insure complete absence of bicarbonate and free carbon dioxide gas. Add 0.3g or 3-4drops of a substance at once. Fit bent delivery tube into another test tube containing lime water, so that the end of delivery tube lie under the surface of the lime water for few minutes, then remove the delivery tube. Close the lime water tube with the thumb and shake vigorously. A white ppt indicates that carbon dioxide has been evolved. The evolution of carbon dioxide indicates the following classes:

1) Carboxylic acids and sulphonic acids but not phenols, so this reaction can be used to distinguish between acids and phenols.
2) Salts of weak bases e.g. aniline sulphate.
3) Nitrophenols, which stronger acid than carboxylic acids.

III.2.6 Action of 30% NaOH solution

It has the same action as in soda lime except the decarboxylation. Place 0.1g of a substance in a test tube; add about 2ml of 30% NaOH solution. Note any reaction in cold then on heating, the reaction mixture boil for few minutes.

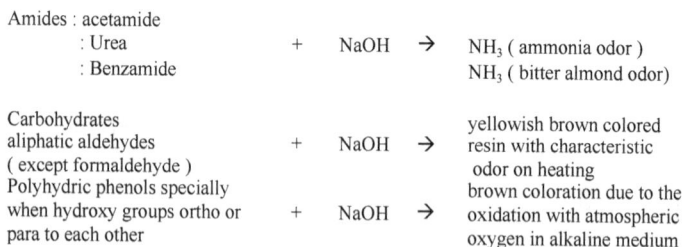

Amides : acetamide : Urea : Benzamide	+	NaOH →	NH_3 (ammonia odor) NH_3 (bitter almond odor)
Carbohydrates aliphatic aldehydes (except formaldehyde)	+	NaOH →	yellowish brown colored resin with characteristic odor on heating
Polyhydric phenols specially when hydroxy groups ortho or para to each other	+	NaOH →	brown coloration due to the oxidation with atmospheric oxygen in alkaline medium

III.2.7 Action of concentrated sulphuric acid

Place 0.1g of the fine powder of the substance in a clean test tube and add gently $\simeq$ 1ml conc. H_2SO_4. Note the status of the reaction (cold or warm) and effervescence or blacking.
- **Carbohydrates and hydroxy acids**
Lactic or tartaric acid + H_2SO_4 $\rightarrow$ blacking and effervescence of CO and/or CO_2
Citric acid + H_2SO_4 $\rightarrow$ yellow and effervescence of CO and/or CO_2 no charring
Polyhydric phenols + H_2SO_4 $\rightarrow$ blacking without effervescence
Benzyl alcohols + H_2SO_4 $\rightarrow$ gelatinous polymer without effervescence

III.2.8 Reaction and coloration with iron (III) chloride

The reagent

Prepare a solution of 2.5% aqueous Iron(III) chloride by dissolving 2.5g of solids $FeCl_3$ in 100ml distilled water, then few drops of conc. HCl may be required to get clear solution. The filtration of the solution may be required to get clear solution.

Procedures

1- In case of free acids, neutral solution of the acid may be prepared by placing 0.2g of acid in a test tube and add excess of ammonium hydroxide solution until the solution is just alkaline. The solution then boiled until the odor of ammonia is completely disappeared.
2- Add few drops of neutral $FeCl_3$ solution to the solution of neutral acid. Coloration/colored ppt is appeared in case of neutral acid solutions, phenols and amines
- Formic, acetic, oxalic and succinic acids give blood-red coloration. On boiling a brown ppt of basic ferric salt will form. On adding HCl, the ppt is dissolved giving a clear solution. Oxalic acid gives no characteristic observation. Succinic acid will give buff colored ppt soluble in dilute HCl.
- Aromatic non-phenolic carboxylic acids will give buff colored ppt soluble in dilute HCl and the corresponding acid will be reprecipitated.
- Phenolic acids (salicylic acid) will give a violet color if dilute solution of the acid where used without neutralization, soluble in dilute HCl.

- Hydroxy aliphatic acids (Lactic, tartaric and citric) will give yellow coloration in neutral solution (a blank test should be tried first).
- Aromatic amines (dissolve in dilute HCl then add few drops of the reagent without neutralization) will give color. Heat the reaction mixture and observe whether the color is changing or remain.
- Phenols (use very dilute aqueous or alcoholic solutions): will give variety of colors, may be obtained according to the number of substitutions of the hydroxyl groups.

III.2.9 Detection of elements (N, S, X) [sodium fusion test for solids only]
$$[CHNSX] + Na / \Delta \rightarrow NaCN + Na_2S + NaX$$

Procedures

Cut a small piece of sodium metal and dry on a filter paper. Place this small piece of sodium in a clean dry small test tube. If the substance is volatile, mix the sample with sodium metal before any heating then proceeds). Heat the tube until sodium melts. Immediately drop the sample directly into the tube (about 10mg of solid or 2-3 drops of liquid). Be sure to drop the sample directly down the center of the tube so that it comes into contact with hot sodium metal. Reheat the tube for some time until it becomes red. Allow the tube to cool to room temperature and then carefully add 10 drops of methanol, a drop at time, to the fusion mixture and stir with glass rod to insure the complete reaction of remaining sodium with CH_3OH. Crush the test tube in 5 -10 ml of distilled water in small beaker. Stir the solution well and then heat to boiling and filter to get clear filtrate.

Test for nitrogen

Ferrous ammonium sulphate solution

Dissolve 5g of ferrous ammonium sulphate in 100ml of distilled water.

Procedures

Use pH paper and a 10% NaOH solution to adjust the pH of about 1ml of the stock solution to pH-13. Add 2 drops of saturated ferrous ammonium sulphate solution and 2 drops of 30% potassium fluoride solution. Boil the mixture for about a minute. Then, acidify the hot solution by adding 30% sulphuric acid dropwise until the iron hydroxide is dissolved. Avoid using of an excess of acid. If nitrogen is present, a dark blue (not green) ppt of Prussian blue $Na_3[Fe (CN)_6]$ will form or the solution assumes dark blue color.

Test for sulphur

Procedures

Acidify about 1ml of the test solution with acetic acid and add a few drops of 1% lead acetate solution. The presence of sulphur is indicated by dark ppt of lead sulphide.

Test for halide

Procedures

Cyanide and sulphide ions interfere with this test. If they are present, they must be removed. To accomplish this, acidify the solution with dilute nitric acid and boil it for about 2 minutes. This will drive off any HCN or H_2S that is formed. When the solution cools, add a few drops of 5% silver nitrate solution. A voluminous ppt indicates the presence of halide: a faint turbidity does not indicate a positive test. Silver chloride is white and readily soluble in conc. NH_4OH, Silver bromide is off-white (quite yellowish) which slightly soluble in NH_4OH, and silver iodide is clear yellow which has poor solubility in NH_4OH.

III.3 Functional group reactions and identification

III.3.1 Alcohols (methyl, ethyl, 1-propanol, 2-propanol, glycerol, benzyl)

Alcohols are neutral compounds. They react with sodium metal and produce H_2 gas. Alcohols usually do not give positive reaction with 2, 4-dinitrophenyl hydrazine test. However, alcohols react with acetyl chloride to form ester and react with Lucas as well, which differentiate them from ester. 1° and 2° alcohols are easily oxidized, whereas 3° alcohols did not oxidize. A combination of the Lucas test and the chromic acid test will differentiate between 1°, 2° and 3° alcohols.

Sodium metal test

Procedures

Cautiously take a few drops of given alcohol into a test tube and add a small piece of sodium metal. Evolutions of gas and heat indicate a positive reaction, according to the following equation:

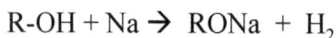

R-OH + Na $\rightarrow$ RONa + H_2

Esterification test

Procedures

Cautiously add about 10-15 drops of acetyl chloride (drop by drop) to about 0.5ml of alcohol in a small test tube. Usually the reaction is exothermic. Therefore, the evolution of heat and gas indicate a positive reaction, according to the following equation: CH_3COCl + ROH $\rightarrow$ CH_3CO_2R + HCl

Oxidation test

Reagents

Chromic acid solutions, dissolves 5g of $K_2Cr_2O_7$ or K_2CrO_4 in 1ml conc. H_2SO_4, and then add 15ml H_2O.

Procedures

1- Dissolve one drop of a liquid or about 10mg of a solid alcohol in 1ml of reagent grade acetone.
2- Add one drop of the chromic acid reagent and note the result that occurs within 2seconds. A positive test for the presence of 1° or a 2° alcohol is the appearance of a blue-green color. 3° alcohols will not give the test within 2 seconds, and the solution will remain orange.
3- To make sure that the acetone solvent is pure and does not give a positive test, add one drop of chromic acid to 1ml of acetone, which does not have an unknown dissolved in it. The orange color of the reagent should persist for at least 3 seconds.

4- If it does not, a new bottle of acetone should be used. This test based on the reduction of Cr(VI) which has orange color into Cr(III) which has green color, when an alcohol is oxidized by the reagent to carboxylic acid (for 1°) or to ketone (for 2°). The reactions occur according to the following equation:

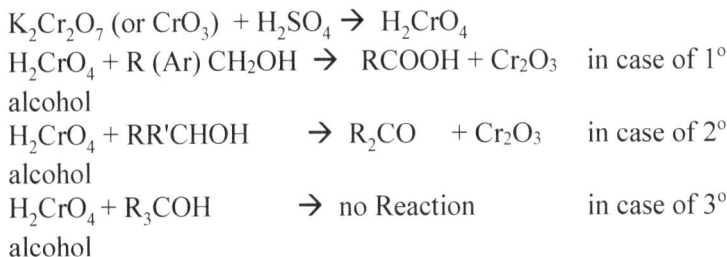

$K_2Cr_2O_7$ (or CrO_3) + H_2SO_4 → H_2CrO_4

H_2CrO_4 + R (Ar) CH_2OH → $RCOOH$ + Cr_2O_3 in case of 1° alcohol

H_2CrO_4 + RR'CHOH   → R_2CO + Cr_2O_3 in case of 2° alcohol

H_2CrO_4 + R_3COH → no Reaction   in case of 3° alcohol

Borax test

Reagent: Dissolve 1g of sodium borate ($Na_2B_4O_7$ in 100ml of water to get 1% solution).

Procedures

Place 5ml of borax solution into a test tube, then add one drop of ph.ph. The pink color will form. Add 1ml glycerol to the solution, the color will disappear. If the mixture is heated, the color will appear again.

Lucas test

Reagent: Cool 10ml conc. HCl in a beaker using ice bath. While still cooling, and with stirring, dissolve 16g of anhydrous zinc chloride in the acid.

Procedures

1- Place 2ml of Lucas reagent in a small test tube and add 3-4 drops of the alcohol. Stopper the test tube and shake it vigorously.
2- Tertiary, benzylic, and allylic alcohols will give immediate cloudiness in the solution as the insoluble alkyl halide separates from the aqueous solution. After a short time, the immiscible alkyl halide will form as separated layer.
3- Secondary alcohol will produce cloudiness after 2-5mins. Primary alcohols will dissolve in the reagent to give clear

solution. Some secondary alcohols may have to be heated slightly to encourage reaction with the reagent.

4- This test will only work for alcohol that soluble in the reagent. This often means that alcohols with more than 6-carbon atoms may not be tested.

$$ROH + HCl / ZnCl_2 \rightarrow RCl + H_2O$$

III.3.2 Aldehydes and ketones

HCHO, chloral hydrate, benzaldehyde, CH_3COCH_3, acetoketone. Aldehydes and ketones are neutral compounds. They react with chromic acid and 2, 4- dinitrophenylhydrazine test and give positive reaction. However, the aldehydic reacts with Tollen's reagent and gives positive test, whereas iodoform test reacts with acetaldehyde and methyl ketones and gives iodoform. Ferric chloride test can be positive in case of those compounds, which have high enol content. The combination of the specific tests can differentiate between them as well between them and the other classes of organic compounds.

2, 4-DNPH test

Reagent

Dissolve 3g of 2, 4-dinitrophenylhydrazine in 15ml of conc. H_2SO_4. Mix 20ml of water and 70ml of 95% ethanol with vigorous stirring, slowly add 2,4-DNPH solution to the aqueous ethanol mixture. After thorough mixing, filter the solution by gravity through a fluted filter paper.

Procedures

Place 2-4 drops of a liquid (or 20 mg if solid) of unknown and dissolve it in a minimum amount of 95% ethanol before adding the reagent), and add 1ml of 2,4-DNPH reagent. Shake the mixture vigorously. Most aldehyde and ketones will give a yellow to red precipitate immediately. However, some compounds will require up to 15min, or even gentle heating, to give a precipitate. The formation of a precipitate indicates a positive test.

Oxidation test

Reagent: The same reagent, which have been used in alcohol.

Procedures: (Follow the same procedures as in alcohol).

A positive test is indicated by the formation of a green ppt and a loss of the orange color in the reagent. With aliphatic aldehyde, the solution turns cloudy within 5sec. and a ppt appears within 30sec. Aromatic aldehydes generally require 30-120sec for the formation of a ppt but some may require even longer. In negative test, usually there will be no ppt. However, in some cases, a ppt will form, but the solution will remain orange in color, according to the following equation:

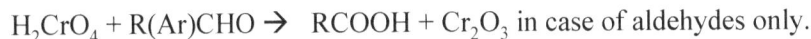

$H_2CrO_4 + R(Ar)CHO \rightarrow RCOOH + Cr_2O_3$ in case of aldehydes only.

Schiff's test

Reagent

If this reagent is not provided, dissolve 1g of fuchsine (resaniline) in 350-400ml of warm water. Cool the solution and pass SO_2 gas through it until the solution is colorless or pale yellow. Dilute to 1000ml. Keep the solution in amber-colored bottle.

Procedures

To 4ml of Schiff's reagent add few drops of the dilute aldehyde solution. Note your observations.

Tollen's test

Reagent

(The reagent must be prepared immediately before the experiments) Solution A, dissolve 3.0 g of silver nitrate in 30ml of distilled water. Solution B; Prepare a 10% sodium hydroxide solution. Mix 1ml of solution A with 1ml of solution B. A precipitate of silver oxide will form. Add enough dilute ammonium hydroxide solution (dropwise) to the mixture to dissolve the ppt.

Procedures

Dissolve 2-4 drops of a liquid (20mg of solid) aldehyde in a minimum amount of dioxane. Add this solution, a little at a time, to the 2 or 3ml of reagent contained in a test tube. Shake the solution well. If a mirror of silver is deposited on the outer wall of the test tube, the test is positive. In some cases, it may be necessary to warm the test tube in a warm water bath. The positive test according to the following equation:

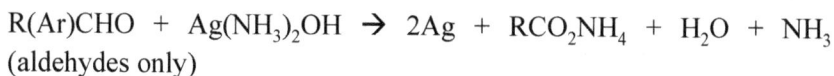

$R(Ar)CHO + Ag(NH_3)_2OH \rightarrow 2Ag + RCO_2NH_4 + H_2O + NH_3$
(aldehydes only)

Iodoform test

Reagent: Dissolve 20g of potassium iodide and 10g of iodine in 100ml water.

Procedures

Dissolve 4 drops of a liquid (or 0.1g of solid) aldehyde in 5ml of dioxane in a large test tube. Add 1ml 10%NaOH solution. Then, add the I_2 - KI solution, drop wise with shaking, until a slight excess gives a definite dark color of I_2.

1- If at this point at least 2ml of the iodine reagent was added and decolourized go to step 3.

2- If less than 2ml if iodine reagent was used, heat the test tube 60° C in a beaker of water. If the solution is decolourized on heating, add more of the iodine solution dropwise until the solution no longer becomes decolourized after 2min of heating at 60° C. Follow this by adding a few drops of NaOH with shaking to decolorize any unreacted iodine remaining in the solution. The iodine will be intensely colored. The solution will remain yellow when it is discharged.

3- Fill the test tube with water and allow it to stand for 15min. A yellow ppt of iodoform will form if the unknown was a methyl ketone or a compound easily oxidized to a methyl ketone. In case of aldehydes, only acetaldehyde will give a positive test. To prove the identity of the yellow ppt as iodoform, collect, dry and determine the melting point of the solid (119°-121°C). The chemical equations as follows:

$$RCHOHCH_3 + NaOH/ I_2 \rightarrow RCOCH_3$$
$$RCOCH_3 + NaOH / I_2 \rightarrow RCOCI_3$$
$$RCOCI_3 + OH^- \rightarrow RCOO^- + CHI_3$$

Fehling's test

Reagent

Solution A, dissolve17.3g of pure copper sulphate ($CuSO_4.5H_2O$) in 180 -200ml of distilled water and dilute the solution to 250ml.
Solution B, dissolve 135g of Rochelle salt and 37g of NaOH in 150ml of distilled water and dilute the solution to 250ml.
To use these solutions (reagent) mix equal volumes of A and B.

Procedures

To 2ml of Fehling's reagent (mixture of A and B) in a clean test tube slowly add 3drops of aldehyde or ketone, Boil gently for 2minutes. If aldehyde is present, the color of the solution will turn into brown that indicates the positive test.

FeCl$_3$ test

Reagent: The preparation of reagent has been mentioned before.

Procedures

Add several drops of a 2.5% aqueous FeCl₃ to a 1ml of a dilute aqueous solution of the unknown (about 1-3% by weight). Some aldehydes and ketones, those that have high enol content, will give a positive ferric chloride test. A positive test indicated by an intense red, blue, purple, or green color. Some colors are transient, and it may be necessary to observe the solution carefully just as the solutions are mixed.

Copper mirror test

Reagent This reagent is a combination of Fehling's reagents and $AgNO_3$ in alkaline medium. So, use the mixture of Fehling's reagent (A and B) and a solution of $AgNO_3$.

Procedures

To 2ml of Fehling's reagent (mixture of A and B) and 1ml of $AgNO_3$ in a clean test tube slowly add 3drops of aldehyde, Boil gently for 2minutes. If aldehyde present, a mirror of copper is deposited on the inner wall of the test tube, the test is positive, which indicate the oxidation of aldehyde to carboxylic acid.

$NaHSO_3$ test

Reagent: Prepare a saturated sodium bisulphite ($NaHSO_3$) solution by dissolving 10g in 50ml distilled water.

Procedures

Shake 1ml of aldehyde with 5ml of freshly prepared $NaHSO_3$. Cool the mixture. Aldehydes and Ketone (which are not branched near the functional group such as acetone or cyclohexanone) will give bisulphite additional compounds. The products are α-hydroxysulphonates which can be crystallized as sodium salts and could be dissolved by the treatment with acids or bases. The reaction occurs as follows:

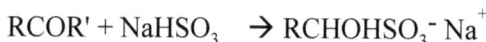

$$RCOR' + NaHSO_3 \rightarrow RCHOHSO_3^- Na^+$$

1- Acetone, cyclohexanone, chloral hydrate will react and give ppt.
2- Formaldehyde and acetaldehyde react with no ppt (water soluble products).
3- Acetophenone and benzophenone will give no reaction.

III.3.3 Carboxylic acids

Formic, acetic, oxalic, succinic, tartaric, citric, benzoic, phthalic, cinnamic, and salicylic and other acids provided. These compounds, which have the formula R(Ar)COOH, are detected mainly by their solubility characteristic. They are soluble in both dilute NaOH and $NaHCO_3$ solutions. Therefore, they will give acidic aqueous solutions which checked by pH by using pH paper. If the compound is an acid, the solution will have a low pH.

NaHCO₃ test

Procedures: Follow the procedure of the action of NaHCO₃ as mentioned before.

Esterification test

Procedures

To 1ml of sample in test tube, add 1ml of alcohol (ethyl alcohol in case of acetic acid and methyl alcohol in case of salicylic acid), then few drops of conc. H_2SO_4. Close and heat the tube for a few minutes and observe the odor of the product.

FeCl₃ test

Reagent: The preparation of reagent has been described before.

Procedures

Follow the same procedure as mentioned before. The acid solution should be neutralized with NH_4OH except salicylic acid. Formic and acetic acids give red -blood color), Oxalic (no color), Tartaric and citric acids give yellow color, succinic, benzoic, phthalic and cinnamic acids give buff color. Salicylic acid gives violet color.

Phthalein test (for phthalic acid):

Reagent: Few grams from phenol.

Procedures

To a small amount of sample in test tube, add few crystals of phenol then 2 drops of conc. H_2SO_4 and heat. Pour the hot mixture in a test tube containing NaOH solution, and then fill the tube with water. A pink color is appeared which decolourized by adding acids.

Fluorescein test (for phthalic acid)

Reagent: Few grams from carboxylic acid, resorcinol, conc. H_2SO_4.

Procedures

To a small amount of sample in test tube, add few crystals of resorcinol solution then 2 drops of conc. $H2SO4$ and heat. Pour the hot mixture in a test tube containing NaOH solution and then fill the tube with water. A green fluorescence color will form.

III.3.4 Carboxylic acids salts

The metal salt of carboxylic acids will give the same results as their acids. Ammonium salts produce ammonia in treatment with 30% NaOH or soda lime. They are all solids and different in solubility, since some carboxylic acids are insoluble in water. Whereas the salts all are soluble in water, they can be treated and decomposed by HCl collected as solids or separated by ether extraction. They can be dissolved in acids and reprecipitated in alkaline medium.

FeCl$_3$ test

Reagent: The preparation of reagent has been described before.

Procedures

Follow the same procedure as mentioned before. Use the solution of the salts without neutralization. Formate and acetate give red -blood color, oxalate (no color), tartrate, citrate give yellow color, succinate, benzoate, phthalate and cinnamate give Buff color. Salicylate gives violet-red color.

Concentrated H_2SO_4 test

Reagent: The preparation of reagent has been described before.

Procedures: Follow the same procedure as mentioned before.

III.3.5 Halogenated acids

Monochloroacetic acid, trichloroacetic acid.

Follow the same experiment for acetic acid except the elemental test will show the presence of halogen.

Beilatein's test

Reagent: Small copper wire.

Procedures

Bend a small loop in the end of a short length of copper wire. Heat the loop end of the wire in a Bunsen burner flame. Cool and dip the wire directly in small compound sample. Heat the wire again, the compound first burn, a green flame will be produced if a halogen is present.

III.3.6 Phenols

Phenol, cresols, resorcinol and catechol, quinol, α-and β-naphthol, salicylic acid.

$FeCl_3$ test

Reagent: The preparation of reagent has been described before.

Procedures

Follow the same procedures as mentioned before. The phenols solution can be used without neutralization. Most phenols give violet - purple color.

Liebermann's test (nitroso test)

Reagent: It has been mentioned before.

Procedures

It is the same principle as in nitrous acid test except sulphuric acid used instead of HCl. Mix the amine sample with sodium nitrite (solid) then add conc. H_2SO_4 after the formation of the diazonium salt, add phenol sample then water and observe the color, then add NaOH solution and note the color in alkaline medium.

Phthalein test (for Phthalate):

Reagent: Phthalic acid, conc. sulphuric acid.

Procedures
To a little amount of sample in the test tube, add a few crystals of phenol then 2 drops of conc. H_2SO_4 and heat. Pour the hot mixture in a test tube containing NaOH solution. A pink color which appeared can be decolourized by adding mineral acids.

Fluorescein test (for phthalate)

Reagent: Phthalic acid, conc. Sulphuric acid.

Procedures To a small amount of sample in test tube, add few crystals of resorcinol then 2 drops of conc. H_2SO_4 and heat. Pour the hot mixture in a test tube containing NaOH. A green fluorescence color will form.

Chloroform test (NaOH and $CHCl_3$)

Reagent: Chloroform, phenols samples and NaOH solution.

Procedures

Take few drops or crystals of phenol in test tube, then add chloroform and NaOH solution and heat the mixture. Note that the red color which is formed is for positive test.

Bromine water test

Reagent: Prepare a saturated solution of bromine in water.

Procedures

Prepare a 1% of aqueous solution of the compound and then add few drops of bromine water drop by drop with shaking, until the color of bromine is no longer discharged. A positive test is indicated by the precipitation of a substitution product at the same time that the bromine color of the reagent is discharged.

III.3.7 Esters

Ethyl acetate, methyl salicylate, aspirin.

Hydroxamate test

Reagent: A mixture of 1ml of hydroxylamine hydrochloride (0.5N) dissolved in 95% ethanol) + 0.2ml NaOH (6N), in addition, 5% FeCl3 solution.

Procedures

Add 2-3 drops of liquid (or 40mg of solid) ester to the 1ml of the reagent. Heat the mixture to boiling for few minutes. Cool the solution and add 2ml of HCl (1N). If the solution is cloudy, add 2ml ethanol. Add a drop of $FeCl_3$ solution (5%). If the color fades, continue add $FeCl_3$ until it persists. A positive test should give a deep burgundy or magenta color, which have enolic characters in acidic medium.

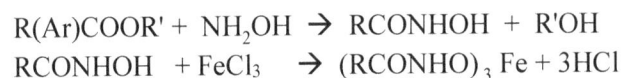

$R(Ar)COOR' + NH_2OH \rightarrow RCONHOH + R'OH$
$RCONHOH + FeCl_3 \rightarrow (RCONHO)_3 Fe + 3HCl$

III.3.8 Amines and salts

Alkyl amines, aniline, p-toluidine, α, β-naphthylamine, diphenylamine, aniline HCl.

Nitrous acid test

Reagent: 5% $NaNO_2$ solution, conc. HCl.

Procedures

To few drops or crystals from the sample add 3-5drops of HCl, then add water to dissolve the test tube content. Add few drops of nitrite solution and note the color of the product. A diazonium salt will form with characteristic color that indicates the amino group in the compound.

Liebermann's test

It is the same principle as in HNO_2 acid test except sulphuric acid used instead of HCl. Mix the amine sample with $NaNO_2$ (solid) then add conc. H_2SO_4. After the formation of the diazonium salt, add phenol or aniline sample then water and note the color, then add NaOH solution. In addition, note the color in alkaline medium.

Hinesburg's test

Reagent: NaOH solution (10%) and p-toluenesulphonyl chloride (or 0.2ml benzene- sulphonyl chloride).

Procedures

Place 0.1ml of liquid (0.1g of solid) amine and 0.2g of p-toluenesulphonyl chloride in a small test tube and then 5ml of NaOH solution. Stopper the test tube tightly, and shake it for 3-5 minutes. Remove the stopper and warm the test tube, with shaking on water bath for one minute. Cool the solution and test a drop with pH paper to see if it is still basic, if not add more NaOH). If a ppt has formed, dilute the basic mixture with 5ml water and shake it well. If the ppt is insoluble, a disubstituted sulphonamide is probably present, which indicates that the unknown was a 2° amine. If no ppt remains after diluting the mixture, or none formed initially, carefully add 5% HCl until the solution is just acidic to litmus paper. If a ppt forms at this point, it should be the monosubstituted sulphonamine, indicating that the original compound was 1° amine. If no reaction occurred during the test, the original compound was probably a 3°.

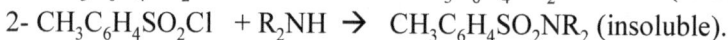

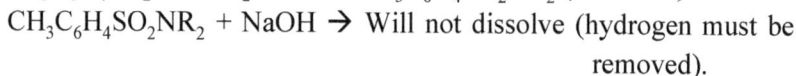

1- $CH_3C_6H_4SO_2Cl + RNH_2 \rightarrow CH_3C_6H_4SO_2NHR$ (soluble in base)

$CH_3C_6H_4SO_2NHR + NaOH \rightarrow CH_3C_6H_4SO_2N^-R\ Na^+$ (soluble salt)

2- $CH_3C_6H_4SO_2Cl + R_2NH \rightarrow CH_3C_6H_4SO_2NR_2$ (insoluble).

$CH_3C_6H_4SO_2NR_2 + NaOH \rightarrow$ Will not dissolve (hydrogen must be removed).

Diazonium test

Reagent: Aniline, Sodium nitrite solution.

Procedures

To 1ml of aniline add 1ml of conc. HCl and add water to get a clear solution, and then add few cold drops of 10% $NaNO_2$. The diazonium salt is formed. Keep the salt in ice bath.

1- To 1st test tube containing one drop of phenol add few drops of NaOH solution then add double of the quantity from the diazonium salt. Azo dye will form and the color depends on the compound used (from brown - reddish-brown-red).

2- Add water to a small amount in 2nd test tube and heat then add $FeCl_3$ (violet color)

3- In 3rd test tube, add to the salt ethanol and heat (intensity increase in color)

III.3.9 Amides and imides

Acetamide, benzamide, succinimide, phthalimide and urea.

30% Sodium hydroxide test

Reagent: The preparation of reagent has been described before.

Procedures

Add 0.2g of the sample to be tested in a test tube containing 3ml 10% NaOH solution. Heat the mixture to the boiling. Note the odor of the ammonia if the amide of form R (Ar)$CONH_2$. Test the vapor with litmus paper (turn red litmus paper into blue).

$R(Ar)CONH_2 + NaOH \rightarrow RCOOH + NH_3$.

FeCl$_3$ test

Reagent: The preparation of reagent has been described before.

Procedures

Follow the same procedure as mentioned before. These compounds will not give any color comparing with amines.

Beurit's test

Reagent: Urea, copper sulphate solution and NaOH solution.

Procedures

In a dry test tube, heat small amount of urea until it melts. Dissolve the product in a small amount of water and try to get clear solution. Add NaOH solution and few drops of copper sulphate and note the purple color, which is formed indicating the formation of a complex from copper ions and the Beurit's compound

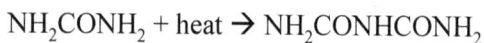

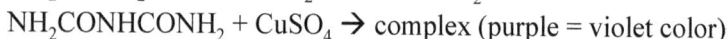

NH_2CONH_2 + heat → $NH_2CONHCONH_2$

$NH_2CONHCONH_2$ + $CuSO_4$ → complex (purple = violet color)

III.3.10 Halogenated hydrocarbon

Chloroform, carbon tetrachloride and benzyl chloride.

Beilatein's test

Reagent: Small copper wire.

Procedures

Bend a small loop at the end of a short length of copper wire. Heat the loop end of the wire in a Bunsen burner flame. Cool and dip the wire directly in small compound sample. Heat the wire again, the compound first burn, a green flame will produce if a halogen is present.

AgNO$_3$ test

Reagent: The preparation of reagent has been described before.

Procedures: They have been described before.

Nitration test

Reagent: The preparation of reagent has been described before.

Procedures: They have been described before.

III.4 Synthesis of some organic compounds

Experiment # 1

Preparation of cyclohexene (dehydration of cyclohexanol)

Introduction
Dehydration of cyclohexanol to cyclohexene can be accomplished by pyrolysis of the cyclic secondary alcohol with an acid catalyst at moderate temperature by distillation over alumina or silica gel. The procedure selected for this experiment involves catalysis by H$_3$PO$_4$. H$_2$SO$_4$ is not more efficient, because charring gives rise to SO$_2$.

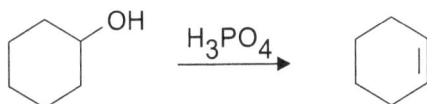

cyclohexanol (B.p = 161°, d = 0.96) cyclohexene (38°, d = 0.81)

Procedures

Assemble a distillation apparatus. Use 100ml distilling flask and a 50ml receiving flask. The collection flask should be immersed to its neck in an ice-water bath to minimize the possibility that cyclohexene vapor will escape into the laboratory. Place 20ml of cyclohexanol and H$_3$PO$_4$(5ml / 85%) in the distilling flask and mix the solution. Add several boiling stones, start circulating the cooling water in the condenser, and heat the mixture until the product begins to be distilled. The temperature of the distilling vapor should be regulated so that it does not exceed $100^{\circ}C$, this can be done if the mixture is heated with Bunsen burner. Hold the burner by its base and apply the heat so as to maintain a slow but steady rate of distillation. If the temperature rises above $100^{\circ}C$ remove the burner for a few seconds before continuing to heat. When, only a few milliliters of residue remains in the distilling flask, stop the distillation. Saturate the distillate with cold NaCl. Add the salt, little by little and shake the flask gently. When no more salt will dissolve, add enough 10% aqueous Na_2CO_3 solutions to make the distilled solution basic. Pour the neutralized mixture into separatory funnel and separate the two layers. Drain the aqueous layer and then pour the upper layer, (cyclohexene) through the neck of the separatory funnel into a 150ml conical flask. Add about 2-3g of anhydrous Na_2SO4 to the flask and swirl occasionally until the solution is dry and clear (about 10-15min). You can stop it here if there is no enough time to finish the experiment. Store the cyclohexene over Na_2SO_4 in a tightly greased stoppered standard - taper flask. Reassemble a distilling apparatus as shown, but this time use a 50ml flask for distillation again. Cool the receiver in an ice water bath. Decant the dry cyclohexene into distilling flask and add a boiling stones. Distill the cyclohexene over range of $80 - 85^{\circ}C$. Weigh the product and calculate the %yield.

Perform the unsaturation tests

1- Place 8 -10 drops of cyclohexanol in two small test tubes, add to one Br_2/CCl_4 and to the second add aqueous solution of potassium permanganate and shake well. Write your observations.

2- Place 8 -10 drops of cyclohexene in two small test tubes, add to one Br_2/CCl_4 and to the second add aqueous solution of potassium permanganate and shake well.

Experiment # 1a

Preparation of 2-methyl-2-butene (dehydration reaction)

Introduction

Amylenes is a generic term applied to the alkenes of formula C_5H_{10}. All of the amylenes are well known, readily available compounds. In this experiment, we will prepare two compounds of this type from common amyl alcohol, 3° amyl alcohol (2-methyl-2-butanol) and 2° amyl alcohol (2-pentanol).

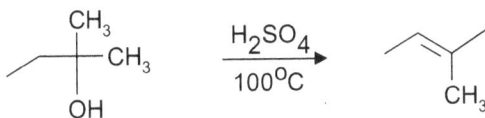

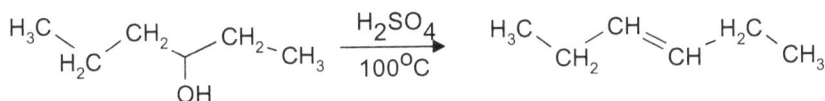

Each of the two dehydration reactions theoretically could proceed in two ways to give two products, but the major product obtained in each case is that represented by the equation. This is in keeping with the general rule that, in the acid-catalyzed dehydration of alcohol which can give rise to two isomeric olefins, the hydrogen atom will be removed from the adjacent (to OH bearing carbon) carbon atom which bears the fewer hydrogen atoms to give the more highly branched olefin. The greater ease of dehydration of 2° vs 3° can be illustrated by the fact that less conc. H_2SO_4 is required to dehydrate 3° than 2° alcohol.

The concerning compounds add bromine vigorously and with the evolution of considerable heat. You should consider this fact in planning an experimental procedure for the addition of Br_2 to either 2-pentene or, 2-methyl-2-butene. The acid-catalyzed addition of H_2O to each of these olefins follows the direction predicted for normal addition of unsymmetrical reagents to unsymmetrical alkenes.

Procedures

Prepare sulphuric acid and water mixture (1:2) by adding cautiously, in small portions, 14ml of conc. sulphuric acid to 27ml of cold water in a 100ml round-bottomed flask. Cool the flask by swirling it gently in an ice bath or a stream of cold water between each addition. Then add 27ml (22g; 0.25mol) of 3° amyl alcohol with cooling and shaking. Mount the flask over a steam bath. (Note 1: for efficient heating, it is advisable to wrap a towel around the flask and extend the towel over the top of the steam bath. Escaping steam is then used effectively to heat the flask) on a ring stand and attach it to an efficient condenser arranged for distillation. Fit the condenser with a curved adapter which led through cotton plug into a 250ml Erlenmeyer receiving flask packed in ice. (Note 2: The compounds are all low boiling points and highly flammable, so efficient consideration is critical). Transfer the cooled product to a small separatory funnel and add 10ml of cold 10% sodium hydroxide solution. Invert the funnel, open the stopcock to release the pressure, then close the stopcock and shake vigorously, stopping occasionally to release the pressure. Tap off and discard the lower aqueous layer and pour the alkenes through the mouth of the separatory funnel into a small dry Erlenmeyer flask. Add about 1g of anhydrous calcium chloride and allow the flask to stand with cooling and occasional shaking. When the hydrocarbon is dry, as indicated by absence of turbidity, transfer it into a small distilling flask fitted with a thermometer, and a dry condenser attached as before to a curved adapter. Distill over a water bath collecting the fraction boiling at 37-43°C in a small tarred collecting bottle packed in ice. Pure 2-methyl-2-butene is reported to boil at 38.5°C at 760mm. Record the data and calculate the % yield. Test the product for unsaturation by the previous methods mentioned.

Experiment # 2

Preparation of cyclohexanone (oxidation of cyclohexanol)

Introduction

Cyclohexanol (B.p = 161.5^{0}C, d = 0.96) can be oxidized either by HNO_3 acid and dichromate or by permanganate. In the case of using HNO_3 acid required a stirring motor and toxic fumes must be disposed furthermore, the product will be adipic acid while using permanganate is free from these limitations, but takes several days, because the permanganate reaction is very slow. Therefore, dichromate in acetic acid is used in the oxidation of cyclohexanol to cyclohexanone. The mechanism of oxidation of cyclohexanol to cyclohexanone by dichromate appears to be the following:

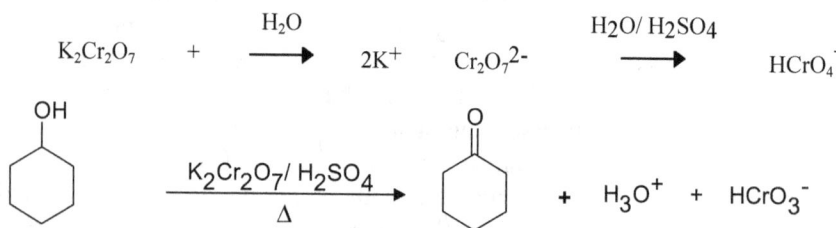

Procedures

In a 125-conical flask dissolve 15g of sodium dichromate dihydrate in 25ml of acetic acid by swirling the mixture on wire gauze above a small burner. The solution, then cool the solution with ice to 15°C. In a second conical flask chill a mixture of 15g of cyclohexanol and 10ml of acetic acid in ice mix the contents of the two flasks after cooling to 15°C, rinse the flask with a little amount of solvent (acetic acid). The exothermic reaction that is soon evident can get out of hand unless controlled. Keep the temperature of the reaction close to 60° by cooling for 15min. No further cooling is needed, but the flask should be swirled occasionally and the temperature, watched for 20-30min. When the temperature begins to drop and the solution become pure green, the reaction is over. Pour the green solution into a 250ml round -bottomed flask, rinse the conical flask with 100ml of water add boiling chips and distill 40ml of liquid, cool the distilling flask slightly, add 40ml of water to it and distill 40ml more. This rather than from an outside source. Cyclohexanone is fairly soluble in water. Dissolving inorganic salts such as potassium carbonate or sodium chloride in the aqueous layer will decrease the solubility of cyclohexanone such that it can be completely extracted with ether and this process is known as "salting out". To salt out the cyclohexanone add to the distillate 0.2g of sodium chloride per ml of water present and swirl to dissolve the salt. Then pour the mixture into a separatory funnel and add 30ml ether, shake and draw off the water layer. Then wash the ether layer with 25ml of 10% sodium hydroxide solution to remove acetic acid and make the solution alkaline. Then draw of the aqueous layer. To dry the ether, which contains dissolved water, shake the ether layer with an equal volume of saturated aqueous sodium chloride solution. Draw off the aqueous layer, pour the ether out of the neck of the separatory funnel into a conical flask add about 5g of anhydrous sodium sulphate for 5min. Remove the drying agent by decantation and evaporate the ether on steam bath under an aspirator tube. Cool the contents to room temperature, Evacuate the crude cyclohexanone under aspirator vacuum to remove final traces of ether weigh the product and find out the yield. The crude cyclohexanone can be purified by simple distillation.

Experiment # 3

Preparation of adipic acid (oxidation of cyclohexene)

Introduction

Cautiously pour 15ml of conc. H_2SO_4 on 15g of ice in a 125ml conical flask, mix thoroughly, then place in an ice bath to cool it to about room temperature. Dissolve 4.5g of sodium dichromate ($Na_2Cr_2O_7.2H_2O$) in 5ml of H_2O Add 2ml of cyclohexene to the cold sulfuric acid mixture. Swirl the contents of the flask to dissolve as much cyclohexene as possible add $Na_2Cr_2O_7.2H_2O$ solution with swirling, by keeping the temperature of the reaction mixture between 26-60°, the flask should not feel very hot or cold to your hand. After the addition is complete the reaction mixture should be green.

Procedures

Place the flask on a steam bath and warm for a few minutes. Pour the contents of the flask into a 100ml beaker containing 15g of ice, then cool the flask in ice until the temperature is below 10°. If no crystals of adipic acid are obtained at this time, scratch the sides of the flask. Filter the crystals by Buchner Funnel and wash the crystals with no more than 2ml of ice water. Recrystallize the product with small amount (5ml) of distilled water. Filter, dry and determine the melting point and calculate the % yield.

Experiment # 3a

Preparation of cyclopentanone (decarboxylation of adipic acid)

Introduction

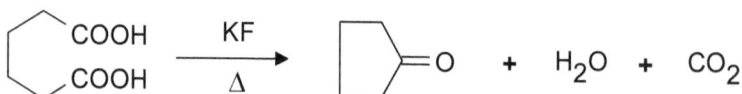

Procedures

Take 50g of adipic acid (0.34mol) and 2g of KF (0.34mol) in dry 200ml round bottom flask, equipped for distillation. Heat the flask gently, until distillation begin and continue until no more distillate is obtained.

Caution: do not overheat, because the overheating will cause charring of the products in the flask, which will make it very difficult to clear.

Transfer the distillate to a separatory funnel and extract with an equal volume of ether. The ether extract is washed three times with 20ml of saturated sodium bicarbonate and once with 20ml of saturated sodium chloride solution. The ether solution should be dried with anhydrous sodium sulphate and distilled over a steam bath to remove the solvent. After ether has been distilled, replace the steam bath with a heating mantle or burner. Collect the cyclopentanone distilling at 120 - 131°C. Weigh and calculate the % yield.

Experiment # 4

Preparation of n-butyl bromide

Introduction

Perhaps the most convenient way to prepare a primary alkyl bromide is to heat the corresponding alcohol with a mixture of sodium or potassium bromide and sulfuric acid. The bromide -sulfuric acid combination serves a double function:

1) It liberates the hydrogen bromide for the reaction and
2) It forms with the alcohol a mixture which boils above the temperature required for fairly rapid reaction of hydrogen bromide with even a primary alcohol.

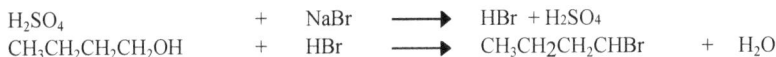

$$H_2SO_4 + NaBr \longrightarrow HBr + H_2SO_4$$
$$CH_3CH_2CH_2CH_2OH + HBr \longrightarrow CH_3CH_2CH_2CHBr + H_2O$$

Procedures

Measure 35ml of water and pour it into 250ml round bottom flask and cautiously add with cooling and careful swirling of the flask, first 35ml of concentrated sulfuric acid and then 37ml (30.0g, 404mol) of n-butyl alcohol. Swirl the mixture and cool by running cold tap water over the outside of the flask. Add in one portion 41.7g (0.30mol) of sodium bromide dihydrate (NaBr.$2H_2O$) construct reflux system and heat gently with small flame while swirling the flask constantly until most of NaBr is dissolved. Then bring the mixture to the reflux temperature and continue gently reflux for 45 min, then set up the apparatus for ordinary distillation with a 250ml Erlenmeyer flask as the receiver. Distill the reaction mixture rapidly until no further droplets of oil can be observed. Pour the total condensate from the receiving flask into a separatory funnel (250ml), properly mounted on a ring stand and tap off n-butyl bromide layer into small conical flask. Slowly add an equal volume of sulfuric acid shake gently. Transfer the mixture to a small separatory funnel (100ml) and add a single drop of water and observe carefully the behavior of the water in order to determine which is n-butyl bromide and which is sulfuric acid layer. Separate the layers and wash the n-butyl bromide, first with 15ml of water, then with 10ml of dilute NaOH solution, and again with 15ml of water. Dry the n-butyl bromide with anhydrous $CaCl_2$. Decant the dry liquid into a small, dry distilling flask; add 2 or 3 small boiling chips. Fit the flask with a thermometer and dry condenser and distill. Collect the distillate which comes over at 99-103°. Weigh the product and calculate your % yield.

Experiment # 4a

Preparation of ethyl iodide

Introduction

Conversion of an alcohol to the corresponding alkyl halide by means of a phosphorus trihalide or a hydrohalic acid, constitute a convenient laboratory synthesis of alkyl halides. Primary, secondary and tertiary alcohols all react readily with phosphorus halides to yield the corresponding alkyl halide. For the synthesis of alkyl halides, phosphorus trichlorides is customarily used directly, but for iodides, it is convenient to generate the corresponding phosphorus triiodide *in situ*, by treatment of phosphorus with iodine.

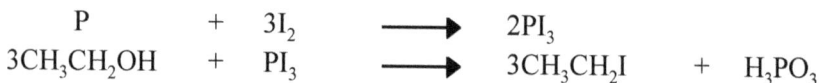

$$P \quad + \quad 3I_2 \quad \longrightarrow \quad 2PI_3$$
$$3CH_3CH_2OH \quad + \quad PI_3 \quad \longrightarrow \quad 3CH_3CH_2I \quad + \quad H_3PO_3$$

Procedures

Place 2.8g (0.088mol) of red phosphorus and 20ml (0.34mol) of ethanol (preferably absolute) in a 200ml flask and fit the flask with a water cool reflux condenser. Detach the flask from the condenser and gradually add 20g (0.08mol) of iodine in portions, of 2-3g each. After each addition of iodine, shake the flask and reattach it to the reflux condenser if necessary to prevent loss of alcohol by evaporation. If the reaction becomes too rapid, cool the flask in cold water. After all the iodine has been added and the mixture no longer heats spontaneously, attach the flask to the condenser and reflux the mixture for 30min. on a water bath to complete the reaction. Remove the flask and cool it in cold water. By means of a distilling head or a bent glass tube connect the flask and the condenser for distillation and distill to dryness. Transfer the crude iodide to a small separatory funnel and add enough 3% NaOH solution so that all of the free iodine is removed by shaking, as evidenced by the discharge of the iodine color. Separate the bottom layer of ethyl iodide from the alkaline solution and then wash with 25ml of water by shaking a mixture of the two in the separatory funnel. Separate the ethyl iodide layer as carefully as possible and dry it in a small flask over 6-10 small granules of anhydrous calcium chloride. During this time wash the condenser, rinse it with a little acetone, and clamp it in a vertical position to dry. Decant the ethyl iodide from the calcium chloride into a small distilling flask and redistill it. A small scale distillation apparatus may be useful here. Carefully note and record the distillation range. Collect the pure product in a dried bottle (previously weighed). Determine its weight and calculate the % yield. Save few ml of ethyl iodide for test.

Experiment # 5

Preparation of acetyl salicylic acid (ASPIRIN)

Introduction

Salicylic acid (o-hydroxybenzoic acid) is a bifunctional compound. It is a phenol (hydroxybenzene) and a carboxylic acid. Either it can undergo two different types of esterification reactions, acting as the alcohol or the acid partner, acetylsalicylic acid (aspirin) is formed, whereas in the presence of excess methanol, the product is methyl salicylate (oil of wintergreen). In this experiment we shall use the former reaction to prepare aspirin. The reaction is complicated because salicylic acid has a carboxyl as well as a phenolic hydroxyl group, and a small amount of polymeric by-product is also formed. Acetylsalicylic acid will react with Sodium bicarbonate to form water -soluble sodium salt, whereas the polymeric byproduct is insoluble in bicarbonate. This difference in behavior will be used to purify the product aspirin.

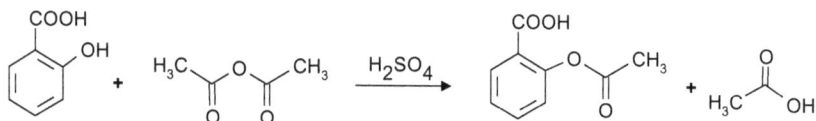

The most likely impurity in the final product is salicylic acid itself, which can arise from incomplete acetylation or from hydrolysis of the product during the isolation steps. This material is removed during the various stages of the purification and the final crystallization of the product. Salicylic acid, like most phenols, forms a highly colored complex with iron(III) ion. Aspirin, which has this group acetylated, will not give the color reaction. Thus, the presence of this impurity in the final product is easily detected.

Procedures

Weigh 5.0g of salicylic acid crystals and place them in a 125ml Erlenmeyer flask. Add 12.5ml of acetic acid anhydride, followed by 10 drops of concentrated sulfuric acid from a dropper and swirl the flask gently until the salicylic acid dissolves. Heat the flask gently on the steam bath for at least 10minutes. Allow the flask to cool to room temperature, during which time the acetylsalicylic acid should begin to crystallize from the reaction mixture. If it does not, scratch the walls of the flask with a glass rod and cool the mixture slightly in an ice bath until crystallization has occurred. After crystals from, add 50ml of water and cool the mixture in an ice bath. Do not add water until crystal formations complete. Usually the product will appear as a solid mass when crystallization has become complete. Collect the product by vacuum filtration on a Buchner funnel. The filtrate can be used to rinse the Erlenmeyer flask repeatedly until all crystals have been collected. Rinse the crystals several times with small portions of cold water. Continue drawing air through the crystals on the Buchner funnel by suction until the crystals are free of solvent. Remove the crystals for air drying. Weigh the crude product, which may contain some unreacted salicylic acid, and calculate the % yield of crude acetyl salicylic acid.

Purification

In 3 test tubes, each containing 5ml of water, dissolve a few crystals of phenol (first tube), salicylic acid (second tube), and your crude product (third tube). Separately add about 10 drops of 1% ferric chloride solution to each tube and note the color. Formation of an iron-phenol complex with Fe(III) gives a definite color ranging from red to violet, depending on the particular phenol present . At the end of the purification procedure you will be asked to repeat this test and note the difference between crude and purified materials. Transfer the crude solid to 150ml beaker and add 45ml of a saturated aqueous sodium bicarbonate solution. Stir until all signs (listen) of reaction have ceased. Filter the solution through a Buchner funnel. If any polymer should be left behind at this point, wash the beaker and funnel with 5-10ml of water. Prepare a mixture of 10ml of concentrated HCl and 25ml of water in a 150ml beaker. Carefully pour the filtrate, a small amount at a time, into this mixture while stirring. The aspirin should precipitate. If it does not, check to see whether the solution is acid, using blue litmus paper. Enough HCl must be added to ensure that the solution is definitely acidic. Cool the mixture in an ice bath , filter the solid by suction, using a Buchner funnel, press the liquid from the crystals with a clean stopper or cork, and wash the crystals well with cold water, It is essential that the water used for this step should be ice-cold. Place the crystals on a watch glass to dry. Weigh the product, determine its melting point and calculate the % yield. Test for the presence of unreacted salicylic acid, using the ferric chloride solution as described above.

Recrystallization

The product prepared according to the above instructions was isolated by precipitation; it should now be prepared as a pure crystalline substance. Water is not a suitable solvent for crystallization because aspirin will partially decompose when heated in water. Dissolve the final product in a minimum amount of hot ethyl acetate (no more than 5-8ml) in a 25ml Erlenmeyer flask, while gently and continuously heating the mixture on a steam bath. When the mixture cools to room temperature, the aspirin should crystallize. If it does not, evaporate some of the ethyl acetate solvent to concentrate the solution and cool the solution in ice water while scratching the inside of the flask with a glass rod (not a fire -polished one). Collect the product by vacuum filtration, using a Buchner funnel. Any remaining material can be rinsed out of the flask with a few ml of cold petroleum ether. Dispose the residual solvents in an appropriate waste container. Submit the crystalline sample in a small vial to your instructor, along with a record of its m.p. Do not forget to test the crystals with $FeCl_3$.

Experiment # 6

Preparation of n-amylacetate by Fischer method (azeotropic esterification)

Introduction

The reaction of a carboxylic acid with an alcohol to produce an ester plus water is known as the Fischer esterification reaction. Mineral acids H_2SO_4 or HCl are usually used as a catalyst. Since the reaction is reversible the law of mass action is applicable .This law states that, when equilibrium is attained in a reversible reaction at constant temperature, the product of the concentration of the reacting substances, each concentration being raised to that power which is the coefficient of the substance in the chemical equation is constant.

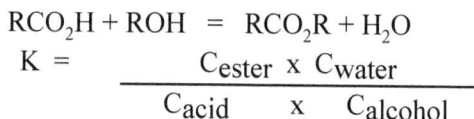

$$RCO_2H + ROH = RCO_2R + H_2O$$

$$K = \frac{C_{ester} \times C_{water}}{C_{acid} \times C_{alcohol}}$$

For most Fischer esterification, k has a value of approximately four (k= 4). Thus, if equal concentrations of alcohol and carboxylic acid are used, the yield of ester, under equilibrium conditions, is about 67%. Thus, yield may be improved either by use of an excess of one the reactants or by removal of water as it is formed (The calculation as mentioned in the text). In the preparation of n-amyl acetate, a binary azeotrope consisting of 91.2% benzene and 8.8% water and boiling at 69.3°, will be formed. The water in the reflux liquid is separated from the benzene by use of a trap of the Dean-stark type.

Procedures

Into the round -bottomed flask place 22g (0.25mol, 27ml) of n-amyl alcohol, 20ml of glacial acetic acid (excess of 0.25mol), 30ml of benzene, and 12-15 drops of conc. H_2SO_4 (as catalyst) along with a few boiling chips. Fill the trap to the level of the side arm with benzene and connect the trap with the flask and condenser. Heat the flask at bath temperature $110°$ which will keep the reacting liquid boiling at such a rate that the liquid condenser drips into the trap at about 100 drops per minute. When the water layer no longer increases in volume, cool the flask for 50-60 min and pour its contents, as well as benzene and water in trap into a separatory funnel .Wash the mixture twice with water (equal volume to the organic layer)to remove acetic and sulfuric acids , separate the benzene layer, and dry it over a little anhydrous $MgSO_4$ while you arrange for fractional distillation . This dried liquid consists chiefly of benzene (b.p. $80°$) and amyl acetate (Bp $148°$). A little amyl alcohol (b.p. $138°$) that escaped esterification may also be present. Distill the liquid, using a 30cm. Packed fractionating column, and collect as n-amyl acetate the portion distilling at $143\text{-}148°$, calculate % yield.

Experiment # 7

Preparation of soap (hydrolysis)

Introduction

Soap can be prepared from animal fats or vegetable oils which make ester with carboxylic acid. They have a high molecules weight and contain the alcohol glycerol. Chemically, these fats and oils are called triglycerides. The principal acids in animal fats and vegetable oils can be prepared from the natural triglyceride by alkaline hydrolysis (saponification). The natural acids are rarely of single type in any given fat or oil. In fact, a single triglyceride molecule in a fat may contain three different acid residues (R-COOH, R'COOH, R"COOH), and not every triglyceride in the substance will be identical. Each fat or oil, however, has a characteristic statistical distribution of the various types of acids possible. The composition of the common fats and oils is given in the essay on fats and oils.

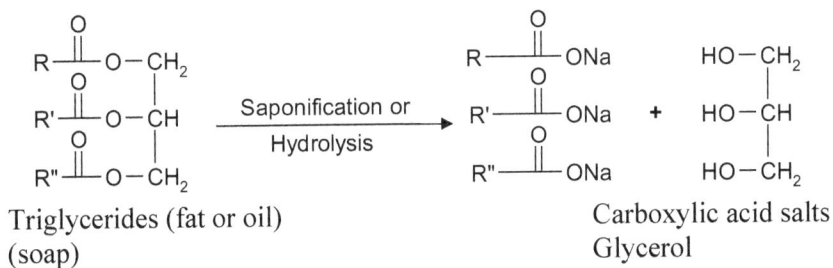

Triglycerides (fat or oil) (soap)   Carboxylic acid salts   Glycerol

The fats and oils that are most common in soap preparations are from animal sources. The length of the hydrocarbon chain and the number of double bonds in the carboxylic acid portion of the fat or oil determine the properties of the resulting soap. For example, a salt of saturated long - chain acid make a harder, more insoluble soap. Chain length also affects solubility. Tallow is the principal fatty material used in making soap. The solid fats of cattle are melted with steam, and the tallow layer formed at the top is removed. Soap - makers usually blend tallow with coconut oil. The coconut oil is added to produce a softer, more soluble soap.

Tallow

Palmitic acid: $CH_3(CH_2)_{14}COOH$, stearic acid: $CH_3(CH_2)_{16}COOH$

Oleic acid: $CH_3(CH_2)_7CH=CH(CH_2)_7COOH$

Coconut oil

Lauric acid: $CH_3 (CH_2)_{10}COOH$, Myristic acid: $CH_3(CH_2)_{12}COOH$

Pure coconut oil yields a soap that is very soluble in water. The soap contains essentially the salt of lauric acid, with some myristic acid. It is so soft (soluble) that it will lather even in seawater. Palm oil contains mainly oleic acid. It is used to prepare Castile soap, named after region in Spain in which it was first made. Toilet soaps generally have been carefully washed free of any alkali remaining from the saponification. As much glycerol as possible is usually left in the soap, the perfumes and medicinal agents are sometimes added. Floating soaps are produced by blowing air into the soap as it is solidified. Soft soaps are made by using KOH, yielding potassium salts rather than the sodium salts of the acids. They are used in shaving cream and liquid soap. Scouring soaps have abrasives and, such as fine sand or pumice.

Procedures

Prepare a solution of 5g of sodium hydroxide dissolved in a mixture or 20ml of water and 20ml of 95% ethanol. Place 10g cooking oil, fat, (a commercial solid shortening works best) in a 250ml beaker and add the solution to it. Heat the mixture on steam bath for at least 45min. Prepare another 40ml of a 50:50 solution of ethanol : water and add it in small portions to the reaction over a 45min period. Stir the mixture constantly. Prepare a solution of 50g of NaCl in 250ml of water in a 400ml beaker. If the solution must be heated to dissolve the salt, it should be cooled before proceeding. Quickly pour the saponification mixture into the cooled salt solution. Stir the mixture thoroughly for several minutes and then cool it to the room temperature in an ice bath. Collect the precipitated soap by vacuum filtration, using a Buchner funnel equipped with fast filter paper. Wash the soap with two portions of ice-cold water. Continue to draw air through the soap to dry the product partially. Allow the soap to dry overnight. Weigh the product. If you are preparing the detergent in above experiment, save a small soap sample for the short comparison test. Submit the remainder of the product to laboratory instructor.

Experiment # 8

Preparation of benzalacetophenone (Aldol condensation)

Introduction

This kind of condensation can occur between acetophenone and benzaldehyde as follows:

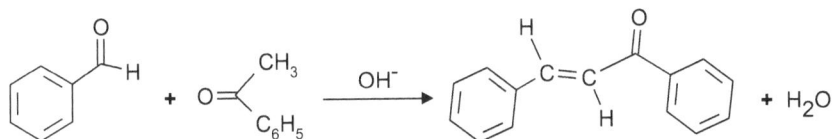

Procedures

Prepare a solution of 4.0g of KOH in 30ml of water in a 250ml conical flask, and add an equal volume of 95% ethanol. Allow the solution to cool to room temperature. While the alkaline solution is cooling, weigh about 12g (0.10mol) of acetophenone and 10.6g (0.01mol) of benzaldehyde These two reactants may also be measured volumetrically with reasonable accuracy. Their densities at 20° are; benzaldehyde= 1.05; acetophenone= 1.03. Add the two carbonyls compounds to the cool alcoholic potassium hydroxide solution and mix the liquids thoroughly by swirling.

Caution: benzalacetophenone is skin irritant, both the solids and its solution should be handled with care. Stopper the flask tightly and shake flask for about 1/2 to 1 hour.

During this time a yellow oil will have formed in the flask. Cool the flask and contents in an ice bath and induce crystallization by scratching the sides of the flask or by seeding. Filter the crystals formed with a Buchener Funnel as soon as they are formed and wash with iced cold alcohol. Recrystallize from 95% ethanol, using 20% excess of the solvent required to dissolve the solid at the boiling point of ethanol. Dry, determine the m.p, weigh and calculate the % yield.

Experiment # 8a

Preparation of dibenzalacetone (1, 5-dipheny-1,4-pentadiene-3-one) (aldol condensation)

Introduction

The reaction of an aldehyde with a ketone employing sodium hydroxide as the base is an example of aldol condensation reaction (the Claisen - Schmidt) Dibenzalacetone is readily prepared by condensation of acetone with two equivalents of benzaldehyde. Benzaldehyde (MW = 106.13 and b.p. = 178°C, d = 104), acetone (MW = 58.08, b.p. = 56°C , d = 0.790). The aldehyde carbonyl is more reactive than that of the ketone and therefore reacts rapidly with the anion of the ketone to give a β-hydroxyketone, which easily undergoes base-catalyzed dehydration. Depending on the relative quantities of the reactants, the reaction can give either mono-or dibenzalacetone. In the present experiment sufficient ethanol, is present as solvent to readily dissolved the starting material (benzaldehyde). In addition, the intermediate (benzalacetone) once formed can then easily react with another mole of benzaldehyde to give the product, dibenzalacetone.

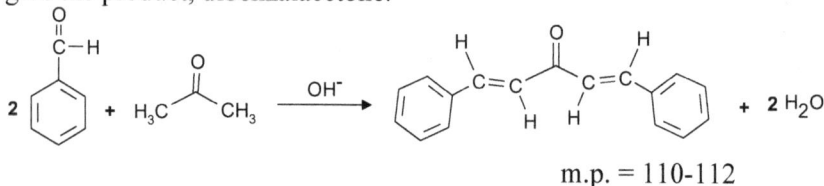

m.p. = 110-112

Procedures

1- Mix 0.50mol of benzaldehyde with the theoretical quantity of acetone, add one-half the mixture to a solution of 5g of NaOH dissolved in 50ml of water and 40ml of ethanol at room temperature ($<25^{\circ}$). After 15min add the reminder of the aldehyde - ketone mixture and rinse the container with a little ethanol to complete the transfer. After a half hour, during which time the mixture is swirled frequently, collect the product by suction filtration on a Buchner funnel.

2- Break the suction and carefully pour 100ml water on the product. Reapply the vacuum. Repeat this process three times in order to remove all traces of NaOH. Finally, press the product as dry as possible on the filter using a cork, and then press it between sheets of filter paper to remove as much water as possible. Save a small sample for melting point determination and then recrystallize the product from ethanol using about 10ml of ethanol for each 4g of dibenzalacetone. Weight the product and calculate % yield and find out the Mp of the pure product.

Experiment # 9

Preparation of aniline (reduction of nitrobenzene)

Introduction

Aniline can be prepared directly from nitrobenzene using Tin metal as reducing agent in presence of conc. HCl.

$$C_6H_5\text{-}NO_2 + Sn/HCl_{(conc.)} \rightarrow C_6H_5\text{-}NH_2$$

Procedures

Before starting the experiment, prepare a bath of ice cold water. Take 15.5g of nitrobenzene and 30 g of granulated tin in a 500ml round bottom flask. Connect a reflux condenser to this flask. Measure 68ml of conc. HCl in a measuring cylinder. Add 10ml of this acid from the top of the condenser and swirl the flask gently. Soon an exothermic reaction will begin. When reacted mixture stops boiling, another 5ml of HCl is added. Keep adding portions of acid and swirling the flask. If the reacted mixture becomes very hot, cool the flask in cold water. After the addition is completed and reaction is subsided, heat the flask on steam bath for 20 minutes with occasional shaking. During this time prepare a solution of NaOH by dissolving 35 g of the alkali in 100ml of water, and cool this solution in a pan of cold water. Remove the reaction flask from the steam bath , cool it in cold water and to this add the

NaOH solution in small portions , Continue swirling the flask during addition of NaOH .After all sodium hydroxide has been added , check the reaction mixture with a litmus paper to see if it is strongly alkaline if not add more sodium hydroxide . Steam distillation can be used for this reaction mixture. Collect the distillate until milky appearance persists. After this collect another 50ml of distillate. Saturate the distillate with NaCl and extract with two portions of 25ml of ether. Combine the ether extracts and dry it with KOH pallets. Decant the ether phase into a distilling flask and remove the ether by distillation on a steam bath. After all the ether has been removed distil aniline on a heating mantle and collect the fraction. Find the weight and calculate % yield.

Experiment # 9a

Preparation of p-nitroaniline

Introduction

The object of this chemical reaction is to perform the electrophilic reaction as well as the application of protection of amine group, then the hydrolysis of amide group. Besides the above crystallization techniques and thin layer chromatography of the resulting compounds will be applied. In this experiment, we convert acetanilide to p-nitro aniline. The sequence of the reaction begins with aniline as shown in the following chemical reaction. The conversion of aniline to acetanilide, the first step, later steps are nitration, hydrolysis and separation of p-nitroaniline from o-nitroaniline by recrystallization.

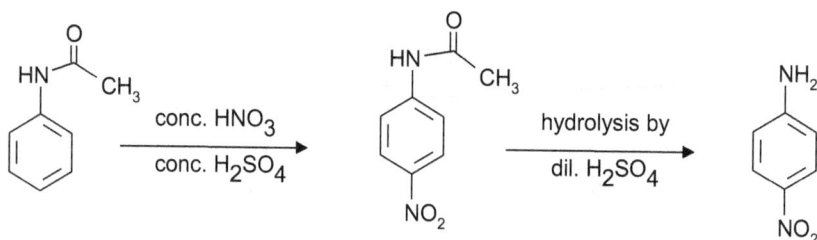

Caution: The nitroanilines (ortho, meta and para) all are toxic. Therefore, avoid the contact with skin, eyes and clothing. Wash all contact areas with large quantities of water.

Procedures

1- Preparation of acetanilide

Place 3ml of aniline in a125ml conical flask. Add approximately 3ml of acetic anhydride dropwise (use hood for this addition). A strong reaction will take place. Keep the flask swirling until the reaction is completed, then add 1ml of water and warm the mixture on a steam bath until all the materials are dissolved. On cooling acetanilide will crystallize. Recrystallize the product from water, using charcoal if the original crystals were dark colored. Isolate the purified material by vacuum filtration and dry it in air. Determine the melting point, weight and the yield.

2- Preparation of p-nitroaniline

1- Place 3.0g of acetanilide in a 125ml Erlenmeyer flask. Add slowly about 5ml of conc. H_2SO_4 to acetanilide. Dissolve most of the solid by swirling and stirring the mixture. Do not be concerned if a small amount of undisclosed solid remains. It will dissolve in later stages of this procedure. Place the flask in an ice bath. Place 1.8ml of conc. HNO_3 in another small flask and add about 5ml of conc. H_2SO_4 to it (prepare this mixture in the hood).

2- Using a disposable capillary pipette, add the mixed acids dropwise to the cooled H_2SO_4 solution of acetanilide. After each addition of acids, swirl the mixtures thoroughly in the ice bath.

3- Do not allow the flask to become warm to touch. After 20min, including the time required for adding the nitric acid-sulphuric acid mixture, add 25ml of an ice-water mixture to the reaction mixture. A suspension of nitroacetanilide isomers will result. Allow this mixture to stand for 5 min., with occasional stirring.

4- To hydrolyze the acetanilide to the corresponding nitroanilines, heat the material in the Erlenmeyer flask, using the dil. H_2SO_4 already present in the flask as hydrolyzing medium. Add a boiling stone and heat the flask using a micro burner. Wire gauze disperses the burner flame.

5- Heat the mixture gently until the solid dissolves, but do not overheat the mixture because the product may decompose. The solution may darken somewhat during this heating period. Cool the flask in an ice bath, and when it is cool, add 30ml of conc. aqueous ammonium hydroxide, in five or six portions, to the material in the flask. This addition must be conducted in the hood, because noxious fumes are evolved during addition. The nitroaniline isomers will precipitate during this addition. Collect the precipitated nitroanilines on Buchner funnel by vacuum filtration. Wash the solid thoroughly with small portions of water (Total =50ml). While continuing the vacuum, allow the solid to dry on Buchner funnel for several minutes.

Crystallization and recrystallization:

Scrape the solid material from the filter paper into 50ml Erlenmeyer flask and add enough hot ethanol to dissolve the solid when the ethanol is boiling. Use the steam bath to heat the solution. When enough ethanol is added to dissolve the solid, while the ethanol is boiling, allow the solution to cool. When the first crystals appear, place the flask in an ice bath to complete the crystallization. Filter the crystals of p-nitro aniline by vacuum filtration using Buchner funnel and a small filter flask. Save the filtrate for a later TLC analysis. Wash the crystals with minimum amount of cold ethanol and allow them to dry by drawing air through them on the filter for a few minutes. Save and dry a small sample of crystals for a later determination of m.p. and TLC analysis. Dissolve the remaining crude p-nitroaniline in ethanol, using 15ml of ethanol for each gram of material. Warm the solution to dissolve the solid. Add 0.5g of activated charcoal to the solution and swirl it for a few minutes. Filter the charcoal from the solution by gravity, using a fluted filter paper. Evaporate some of the solvent to one third of its volume by using steam bath or hot plate. Allow the solution to cool. When the first crystals appear, place the flask in an ice bath. After the crystallization is completed collect the solid material by filtration as before. Dry the solid crystals. Weight the product and calculate % yield and find out the M.p of the pure product.

TLC analysis

Dissolve samples from crude p-nitroaniline and purified product each in few drops of C_2H_5OH. Each sample is spotted on the TLC plate, using CH_2Cl_2 as a solvent for this technique.

Experiment # 10

Preparation of ethyl p-aminobenzoate (Bezocaine)

Introduction

In this experiment a procedure is given for the preparation of local anesthetic benzocaine, by the direct esterification of p-aminobenzoic acid with ethanol (It of optional to test the prepared anesthetic on a frog's leg muscle). The aim of the reactions is to prepare Bezocaine (as drug) from p-aminobenzoic acid (as vitamin for bacteria). The synthesis of these compounds involves in the following chemical equations which are outlined as:

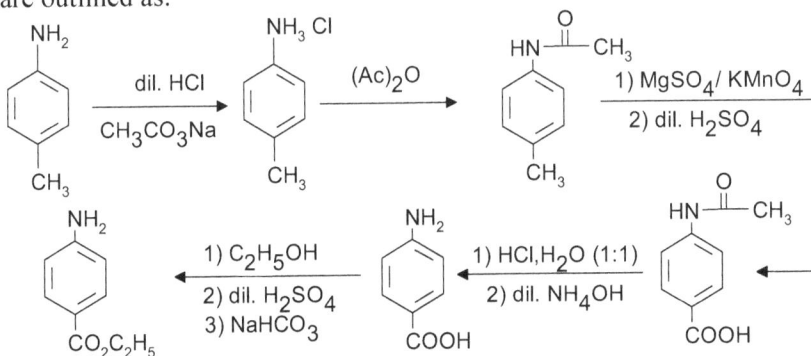

1- The conversion of the commercially available p-toluidine into the corresponding amid N-acetyl-p-toluidine (or p-acetotoluidine). The oxidation of the methyl group with $KMnO_4$ gives carboxylic acid group.

2- The hydrolysis of the amide functional group to remove the protecting acetyl group to produce p-amino-benzoic acid. Then the direct esterification of p-aminobenzoic acid with ethanol to get the ester form.

Procedures

1) N-acetyl-p-toluidine:

Place 8g of powdered p-toluidine in a 500ml Erlenmeyer flask. Add 200ml of water and 8ml of conc. HCl. If necessary, warm the mixture on a steam bath, with stirring, to facilitate solution. If the solution is dark, add 0.5 -1.0g of decolorizing charcoal, stir it for several minutes, and filter by gravity. Prepare a solution of 12g of sodium acetate trihydrate in 20ml of water. If necessary, use a steam bath to warm the solution until all the solid is dissolved.

Warm the decolorized solution of p-toluidine hydrochloride to 50°C. Add 8.4ml of acetic acid anhydride (d=1.08g/l), stir rapidly, and immediately add the previously prepared sodium acetate solution. Mix the solution thoroughly, and cool the mixture in an ice bath. A white solid should appear at this point. Filter the mixture by vacuum, using Buchner funnel, wash the crystals with cold water 3 times, and allow the crystals to stand in the filter to air dry while the vacuum is maintained. These crystals will not be isolated and dried but used directly in the next preparation.

2) p-acetamidobenzoic acid

Place the previously prepared, wet N-acetyl-p-toluidine in a 1L beaker, along with 25g of $MgSO_4.xH_2O$ and 350ml H_2O. Place the flask on a steam bath, and adjust the steam flow to a gentle rate. Add sludge made up to 30g $KMnO_4$ and a small amount of H_2O to the reaction in approximately tea-spoon quantities. Increase the flow rate of the steam and allow the reaction to proceed for one hour. It is important to heat the reaction mixture thoroughly. It is necessary to place the beaker down in the steam cone to heat the beaker effectively.

During the reaction period, the mixture must be stirred every few minutes. After one hour the mixture should be quite brown. Vacuum filter the hot solution through a bed of celite. Wash the precipitated MnO_2 with a small amount of hot water. If the filtrate shows the presence of excess permanganate by its purple color, add not more than 1ml of ethanol, heat the solution in the steam bath for another 30min, and filter the hot solution through fluted filter paper once more. Cool the colorless filtrate and acidify with excess 20% H_2SO_4 solution. A white solid should form at this point. Filter the solid by vacuum and dry it in oven. The yield based on p-toluidine and the Mp should be determined at this stage.

3) p-aminobenzoic acid (PABA)

Prepare a solution of HCl by mixing 24ml of conc. HCl and 24ml of water. Place the previously prepared p Acetamidobenzoic acid in 250ml round-bottomed flask that has a reflux condenser attached to it. Add the HCl solution and heat the mixture gently under reflux for 30min. Allow the reaction mixture to cool, transfer it to 250ml Erlenmeyer flask, add 48ml of water, and make the reaction mixture just alkaline (pH= 8-9, use pH paper) with dilute ammonium hydroxide solution. For each 30ml of the final solution, add 1ml of glacial acetic acid, chill the solution in an ice bath, and initiate the crystallization, if necessary by scratching the wall of the flask with glass rod. Filter the crystals by vacuum and allow the air pass through to insure the drying of the solid. Weight the product and calculate %yield and this step, basing the overall yield on p-toluidine. Determine the m.p. of the pure product. (m.p. of pure p-aminobenzoic acid is 186 – 187°C. Frequently the m.p. of the product is somewhat lower. Recrystallization is not necessary before the p-aminobenzoic acid is used in the next experiment.

4) Ethyl p-aminobenzoate (Bezocaine)

Place 5.0g of p-aminobenzoic acid which previously prepared in 250ml round-bottomed flask. Add the 65ml of 95% ethanol, swirling gently to help dissolve the solid (not all the solid will dissolve). Cool the mixture in an ice bath and slowly add 5ml of conc. H_2SO_4. A large amount of precipitate will form when the H_2SO_4 is added, but this solid will slowly dissolve during the reflux that follows. Heat the mixture gently under reflux on a steam bath for 2hrs. The flask may have to be placed down in the steam cone to obtain the proper reflux rate. Swirl the contents of the flask at approximately 15min intervals during the first hour of the reflux. Transfer the contents to 400ml beaker, add in small portion a 10% Na_2CO_3 solution (60ml needed), to neutralize the mixture. After each addition of Na_2CO_3 solution, extensive gas evolution will be perceptible until the mixture is nearly neutralized. When gas is no longer evolved as you add a portion of Na_2CO_3, check the pH of the solution and add further portions of Na_2CO_3 until the pH=9 or above. Decant the aqueous mixture away from any solid formed from neutralization; pour the mixture into a separatory funnel (250ml). Add a100ml ether to the separatory funnel and shake vigorously. Separate the mixture and save the upper ether layer. Dry it with $MgSO_4$ (about 2 spatulafuls), then gravity- filter the ether to remove the drying agent, and remove the ether and ethanol by evaporating them over hot plate in hood. When most of the solvent has been removed (about 5ml remaining), the oil will be visible in the flask. Add hot 95% ethanol and heat the mixture on the hot plate until the oil dissolves. Add water to the ethanol solution until the oil appears again (extensive cloudiness) and then cool the mixture in an ice bath. The oil that may form initially will crystallize during the swirling in the ice bath. Collect the benzocaine by vacuum filtration, using Buchner funnel. After the solid has been dried overnight at room temperature on a piece of filter paper, then determine the weight of the product and calculate %yield. Determine the M. p of your pure product. (M.p of pure benzocaine is 92°C).

Experiment # 11

Preparation of p-chlorotoluene (Sandmeyer reaction)

Introduction
The Sandmeyer reaction is a versatile means of replacing an aromatic amine group with a halide or cyanide group through reaction of the copper (I) halide or potassium cyanide with a diazonium salt. P-toluidine is dissolved in the required amount of HCl, two more equivalents of acid are added, and the mixture cooled in ice to produce a paste of the crystalline amine hydrochloride. When this salt is treated at $0\text{-}5^{\circ}$C with one equivalent of Sodium nitrite, nitrous acid is liberated and reacts to produce the diazonium salt. The excess HCl beyond the two equivalents required to form the amine hydrochloride and react with sodium nitrite maintains acidity sufficient to prevent formation of the diazoamino compound and rearrangement of the diazonium salt. Copper (I) chloride is made by reaction of $CuSO_4$ with Na_2SO_3 (which is produced as required from the cheaper sodium bisulphate). The white solid is left covered with the reducing solution for protection against air oxidation until it is to be used and then dissolved in HCl. On addition of the diazonium salt solution, a complex mixture forms and rapidly decomposes to give p-chlorotoluene and nitrogen. The mixture is very colored, but steam distillation leaves most of the impurities and all salts behind and gives material substantially pure except for the presence of a trace of yellow pigment which can be eliminated by distillation of the dried oil.

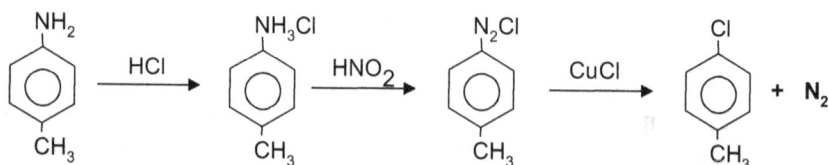

$$2CuSO_4.5H_2O + 4NaCl + NaHSO_3 + NaOH \rightarrow CuCl + 3Na_2SO_4 + 2HCl + 10H_2O$$

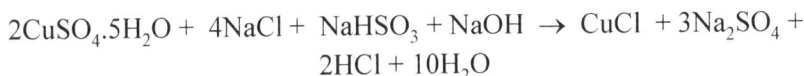

249.71   58.45   104.97   40.01   99.02

Procedures

1- Copper (I) chloride solution

In a 500ml round-bottom flask (to be used later for steam distillation) dissolve 30g of copper (II) sulphate crystals ($CuSO_4.5H_2O$) in 100ml of water by boiling and then add 10g of NaCl, which may give a small precipitate of basic $CuCl_2$. Prepare a solution of Na_2SO_3 from 7g of sodium bisulphite, 4.5g of NaOH, and 50ml of H_2O and add this (not too rapidly) to the hot $CuSO_4$ solution (rinse flask and neck). Shake well and put the flask in a pan of cold water in a slanting position favorable for decantation and let the mixture stand to cool and settle during the diazotization. When you are ready to use $CuCl_2$, decant the supernatant liquid, wash the white solid once with water by decantation, and dissolve the solid in 45ml of conc.HCl .The solution is susceptible to air oxidation and should not stand for an appreciable time before use.

2- Diazotization

Put 11.0g of p-toluidine and 15ml of H_2O in a 125ml Erlenmeyer flask. Measure 25ml of conc. HCl and add 10ml of it to the flask. Heat over a free flame and swirl to dissolve the amine and hence ensure that it is all converted into the hydrochloride. Add the remaining acid and cool thoroughly in an ice bath and let the flask stand in the bath while preparing a solution of 7g of sodium nitrite in 20ml of water .To maintain a temperature of 0-5°C during diazotization , add a few pieces of ice to the amine hydrochloride suspension and add more later as the first ones melt . Pour it in the nitrite solution in portions during intervals with swirling in the ice bath. The solid should dissolve itself to a clear solution of the diazonium salt. After 3-4min test it for excess nitrous acid by dipping a stirring rod in the solution and touch off the drop on the wall of the flask, put the rod in a small test tube, and add a few drops of H_2O. Insert a strip of starch-iodide paper. An instantaneous deep blue color is due to a starch-iodine complex which indicates the presence of nitrous acid (The sample is tested by diluted with H_2O

because strong HCl alone produces the same color on starch-iodide paper after a slight induction period). Leave the solution in the ice bath.

3- Sandmeyer reaction

Complete the preparation of copper (I) chloride solution, cool it in the ice bath, pour it in the solution of diazonium chloride through a long - stemmed funnel, and rinse the flask. Swirl it occasionally at room temperature for 10min and observe initial separation of a complex of the two components and its decomposition with liberation of nitrogen and separation of oil. Arrange it for steam distillation or general generation of steam in situ by simply boiling the contents of the flask over a flame using the apparatus for simple distillation. Add more water during the distillation. Do not start the distillation until bubbling in the mixture has practically ceased and oily layer is separated. Then steam it distill it and note that p-chlorotoluene, although lighter than the solution of inorganic salts in which it was produced is heavier (density 1.07) than water. Extract the distillate with a little ether. Wash the extract with 10% sodium hydroxide solution to remove any p-cresol present, then wash with saturated sodium chloride solution; dry the ether solution over about 5g of anhydrous sodium sulphate and filter or decant it into a tarred flask. Evaporate the ether and determine the yield and % yield of product. Your yield should be about 9g.

Experiment # 12

Preparation of 3-aminobenzoic acid (nitration of methyl benzoate, reduction and hydrolysis)

Introduction

C_6H_6 and somewhat less reactive aromatic compounds such as methyl benzoate can be nitrated with a mixture of HNO_3/H_2SO_4 that ionizes completely to generate the nitronium and hydronium ions. A strongly deactivated benzene ring such as found in nitrobenzene requires the use of heat, conc. H_2SO_4, and fuming HNO_3 for nitration of m-dinitrobenzene and 1,3,5-Trinitrobenzene cannot be prepared by nitration of m-dinitrobenzene. In the present experiment sulfuric acid serves as the solvent: and nitration occurs at the meta position because of the partial positive charge residing at the o- and p-positions: Nitration of the aromatic ring affords entry numerous derivatives. The nitro group is easily reduced to the amine, e.g. by using hydrogen and a catalyst or Sn, Fe and Zn and HCl.

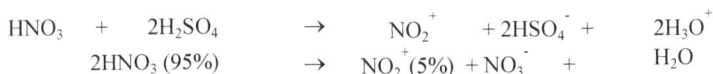

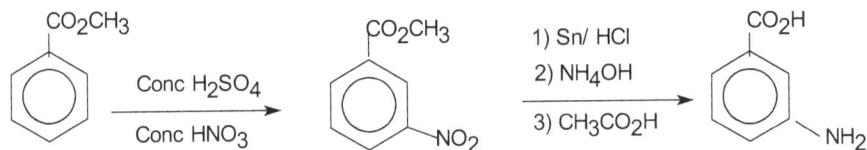

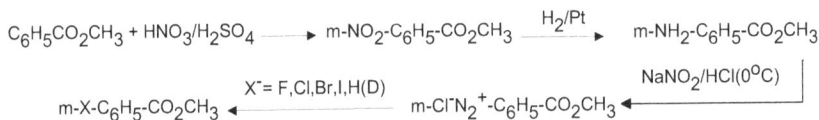

$$HNO_3 \quad + \quad 2H_2SO_4 \quad \rightarrow \quad NO_2^+ \quad + 2HSO_4^- + \quad 2H_3O^+$$
$$2HNO_3 \, (95\%) \quad \rightarrow \quad NO_2^+ \, (5\%) + NO_3^- \quad + \quad H_2O$$

With neutral or basic reducing agents aromatic nitro compounds can be reduced in a stepwise fashion to nitroso, hydroxylamine, azo and hydrazo derivatives on the way to the most highly reduced from, the amine. With metal and acid reduction the reaction is thought to have been involved a series of electron and proton transfers. Advantage can be taken of the fact that the amine once formed is a powerful o-, p-director and activator of the aromatic ring in carrying out further aromatic substitution reactions. The amine can be diazotized and converted via Sandmeyer-type reactions, into a host of derivatives. Using methyl benzoate as an example the following can be prepared easily. During the course of the present reaction not only does the nitro group get reduced but also the ester is hydrolyzed to the carboxylic acid at the same time, because the condition of refluxing aqueous acid is normally used for hydrolysis. In the present nitration experiment the m-nitro substituted methyl benzoate can be isolated in greater than 80% yield. Washing the crude crystalline product with cold methanol to remove the oil consisting of the o-nitro product, some of the desired meta product, and a dinitro ester. Unchanged methyl benzoate has been detected.

1- Preparation of methyl benzoate

In a 500ml round-bottomed flask place a mixture of 30g (0.246mol) of benzoic acid, 80g (101ml, 2.5mol) of absolute methanol and 5g (2.7ml) of conc. H_2SO_4. Add a few small chips of porous porcelain attach a reflux condenser and boil the mixture gently for 4hrs (1). Distill off the excess of alcohol on a water bath (rotary evaporator) and allow it for cooling. Pour the residue into about 250ml of water contained in a separatory funnel and rinse the flask with a few ml of water which is also poured into the separatory funnel. If owing to the comparatively slight difference between the density of the ester and of water, difficulty is experienced in obtaining sharp separation of the lower ester layer and water, add 10-15ml of carbon tetrachloride (2) and shake the mixture in the funnel vigorously; upon standing, the heavy solution of methyl benzoate in the carbon tetrachloride separates sharply and rapidly at the bottom of the separatory funnel. Run off the lower layer carefully, reject the upper aqueous layer, return the methyl benzoate to the funnel and shake it with a strong solution of $NaHCO_3$ until all free acid is removed and no further evolution of CO_2 occurs. Wash once with water, and dry it by pouring it into a small dry conical flask containing about 5g of $MgSO_4$. Stopper the flask, shake for 5mins and allow standing for at least half an hour with occasional shaking. Filter the methyl benzoate solution through a small fluted filter paper directly into a round-bottomed flask fitted with still-head carrying a 360°C thermometer and an air condenser. Add a few boiling chips and distill from an air bath; raise the temperature slowly at first until all CCl_4 is passed over and then heat more strongly. Collect the methyl benzoate (a colorless liquid) at 198-200°C. The yield is 31g (92%).

Notes:

1- Slightly improved results may be obtained by increasing the time of heating.
2- Alternatively, the ester may be extracted with two 50ml portions of ether.

The ethereal solution is washed with concentrated $NaHCO_3$ solution (handle the separatory funnel cautiously as carbon dioxide is evolved) until effervescence ceases, then with water, and dried over $MgSO_4$. The ether is then removed by flash distillation and the residual ester distilled.

2- Nitration of methyl benzoate

Procedures
[Methyl benzoate (M.W. =136.16, b.p. =199.6°C, d =1.09), Methyl-3-nitrobenzoate (M.W. =181.15, m.p. =78°C)]

In a 125ml Erlenmeyer flask cool 12ml of conc. H_2SO_4 to 0°C and then add 6.1g of methyl benzoate. Again cool the mixture from 0-10°C. Now add dropwise, using a Pasteur pipette, a cooled mixture of 4ml of conc. H_2SO_4 and 4ml of conc. HNO_3. During the addition of the acids, swirl the mixture frequently and maintain the temperature of the reaction mixture in the range of 5-15°C. When all the HNO_3 has been added, warm the mixture to room temperature and after 15min. Pour it on 50g of cracked ice in a 250ml beaker. Isolate the solid product by suction filtration using a small Buchner funnel and wash well with water, then with two 10ml portions of ice-cold methanol. A small sample is saved for a melting point determination. The remainder is weighed and crystallized from an equal weight of methanol. The crude product should be obtained in about 80% yield and with a m.p. of 74 -76°C. The recrystallized product should have m.p. of 78°C.

3- Reduction and hydrolysis of methyl 3-nitrobenzoate

[Methyl 3-nitrobenzoate (M.W.=181.15, m.p. =78°C)
3-Amino benzoate (M.W.=137.14, m.p. = 174°C)]

Normally the basic product from a reduction like this is isolated from the diluted reaction mixture by addition of enough NaOH to neutralize the acid and dissolve the tin in the form of sodium stannite. In the present experiment this procedure will not work. The reaction mixture is made weak by basic product using ammonium hydroxide, which causes amphoteric stannous oxide to precipitate. This salt is removed by filtration and the solution is brought to a pH=5 with CH_3CO_2H at which point the product will have minimum solubility in H_2O. The product is the carboxylic acid and not the ester; the reaction conditions, refluxing HCl, are those commonly used to hydrolyze esters.

Procedures

Place 4.0g of methyl 3-nitrobenzoate, 9g of mossy Tin, and 20ml of conc. HCl in a 100ml round bottomed flask fitted with a reflux condenser. A trap for HCl vapors is connected to the top of the condenser with a rubber stopper, a short length of glass tubing , and rubber tubing leading to a filter flask .The filter flask is half filled with water and is fitted with a glass tube inserted through a stopper to within 1 cm of the surface of the water. Heat the flask with a flame until the reaction begins, and shake the reaction mixture every 2 or 3min by moving the entire apparatus, ring stand and all, "back and forth". Heat it from time to time as the reaction slackens. At first the crystalline nitro ester will be seen floating on the surface of the acid. As the reaction proceeds, yellow oil replaces the crystals, and finally the oil disappears to give a water-clear solution. After most of the tin is gone and the reaction is complete (about 25min), decant the hot solution (in the hood) from any residual tin into a 125ml Erlenmeyer flask and rinse the reaction flask with a few ml of water. Make the solution alkaline pH = 8 with concentrated ammonium hydroxide. Remove the precipitate of the salt by filtration 50mm Buchner funnels. Use the filtrate to rinse out the Erlenmeyer flask and the filter cake of tin salts with a little water. Acidify the clear filtrate with glacial acetic acid to pH ~ 5 (use pH paper), cool in ice, and collect the pure white granular crystals by suction filtration. If no product separates, evaporate the solution to a volume of ~ 30ml on the rotary evaporate. Wash the filter cake with a few milliliters of ice-cold water and press the product as dry as possible, allow it to air dry for a day or so, and determine the % yield and m.p.

Experiment # 13

Preparation of 3-methyl-1-phenyl-5-pyrazolone

Introduction

Starting from this experiment we will deal with heterocyclic compounds. The interest in phenyl methyl pyrazolone arises as it is an intermediate in the synthesis of antipyrine, a compound that Knorr, as early as 1883, found which has strong febrifuge action. The preparation of phenyl methyl pyrazolone also serves as a prototype of syntheses in which a fairly complex heterocyclic compound is prepared by condensation of a bi- or poly-functional compound (such as ethyl acetoacetate) with a readily available organic source of the hetero atom (such as phenyl hydrazine. The Skraup synthesis of quinolines, the Hantzsch synthesis of pyridines and the Knorr synthesis of pyrroles are but a few examples of useful preparative methods which fit this general pattern.

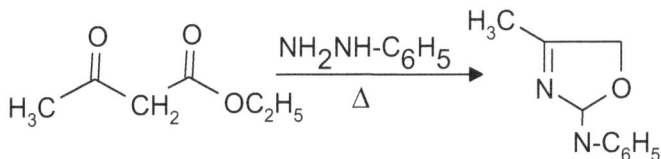

Procedures

In large test tube mix 6.5g (6.35ml, 0.5mol) of ethyl acetoacetate with 5.4g (4.9ml, 0.5mol) of phenyl hydrazine. Note that a little heat is evolved and that the original clear solution becomes turbid owing to the separation of droplets of water in the formation of the phenyl hydrazine of ethyl acetoacetate. Fit the tube with a cork stopper through which is inserted a piece of glass tubing about 18 inches long. Heat the tube in a bath of molten paraffin wax (or other type of bath for at a temperature of 135 -145°. When the heating period has been completed, pour the contents of the tube into a small beaker, cool the beaker in cold water and stir the material with 30-40ml of ether. As soon as crystals form, chill the beaker in ice and collect the product on the Buchner funnel. Recrystallize the phenyl methyl pyrazolone from alcohol or hot water. Weight the pure product and calculate % yield and find m.p.

Alternative procedures

In 100ml bottomed head flask, heat a mixture of ethyl aceto-acetate (6.5g; 6.35ml) and freshly distilled phenyl hydrazine (5.4g ; 4.9ml) at 120°C (oil bath, temp. 120-130°) for one hour. Cool the resulting red oil and stir with ether (50ml) until solidification occurs then filter off the crude product. 3-Methyl-1-phenyl-5-pyrazolone crystallizes from 50% aqueous ethanol (3ml/g) as colorless needles: m.p. = 125-127° and yield = 6.5 - 7.5 g (75 - 86%).

Experiment # 14

Preparation of barbituric acid (hexahydro-2,4,6-trioxopyrimidine)

Introduction

Barbituric acid is prepared by condensing diethyl malonate or its derivatives, with urea in the presence of a base. Interaction of diethyl malonate with urea in the presence of sodium ethoxide yields hexahydro-2, 4,6-trioxo-pyrimidine.

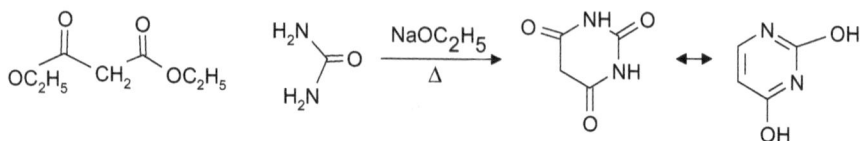

Procedures

In a 250ml round bottomed flask, dissolve freshly prepared sodium (1.15g) in ethanol (25ml). To the solution add successively, diethyl malonate (8g: 7.9ml) and a solution urea (3g in hot 25ml ethanol). Heat the resulting mixture on a water-bath at 100°C for 1½ hr, during which time the sodium salt of the product precipitated. Filter the mixture, and dissolve the residual sodium salt in water (100ml). Acidify the resulting solution with conc. HCl and filter off the crude product. Hexahydro-2, 4, 6-trioxo-pyrimidine crystallized from water (8ml/g) as colorless prism of the dehydrate, which decomposes on heating. Yield = 3.8 - 4g (59.5 - 62.5%).

Experiment # 15

Preparation of 2,5-dioxopiperazine

Introduction

2, 5-Dioxopiperazine results from the action of heat on certain α-amino acids. Heat treatment of glycine (eg. in boiling glycerol or ethylene glycol) leads to the formation of 2,5-Dioxopiperazine.

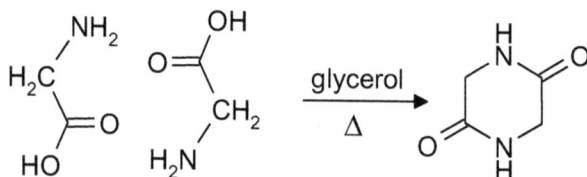

Procedures

In a 250ml round bottomed flask, heat a solution of glycine (15g) in ethylene glycol (90ml) under reflux for ½hr. Cool the dark-brown solution in ice-water bath and filter off the precipitate crude product. 2, 5-Dioxopiperazine crystallized from water (6ml/g) (decolorizing carbon) as white plates m.p. >300°C at which it will decomposes. Yield = 4 -5g (35 - 44 %).

Experiment # 16

Preparation of quinoline (Skraup synthesis)

Introduction

The reactions for the preparation of quinoline can be outlined as under in which the preparation should be carried out in fume cupboard.

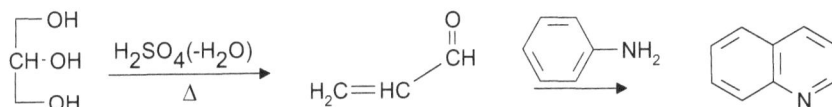

Procedures

Equip a 250ml three-necked flask with a double surface condenser, a sealed stirrer and a screw-capped adapter carrying a thermometer positioned so that subsequently the temperature of the reaction mixture may be noted.

Place 10g (9.8ml; 0.107mol) of pure aniline, 15g (0.163mol) of glycerol and 0.5g of iodine in the flask. Stir the reaction mixture and add down the condenser from a dropping funnel 30g (16.4ml; 0.306mol) of conc. H_2SO_4. Reaction soon commences, the temperature rises to 100-105°C. Heat the flask gradually, with stirring it in an air bath or oil bath to 140°C. The reaction proceeds with the evolution of SO_2 and a little iodine vapor and the liquid reflux. Continue heating at 170°C for about one hour, allow it to cool itself and then add cautiously with stirring sufficient 5M NaOH solution (about 85ml) to render the mixture alkaline. Rearrange the apparatus for distillation and steam distill until no more oily drops pass over. The distillate contains quinoline and a little aniline. Extract the distillate with three 25ml portions of ether, combine the ether layers extract and remove the ether on a rotary evaporator. To remove the aniline present in the residual crude quinoline, advantage is taken of the fact that bis-quinolinium tetrachlorozincate(II) $[(C_9H_8N)_2^+][(ZnCl_4)^{2-}]$ is almost insoluble in H_2O and crystallizes out, while under the experimental conditions, π-anilinium tetrachlorozincate(II) $[(C_6H_5NH_3)^+][(ZnCl_4)^{2-}]$ remains in solution. Dissolve the crude quinoline in 100ml dil HCl (1:4 by volume), warm the solution to $60^{\circ}C$ and add, with stirring a solution of 13g (0.095mol) of $ZnCl_2$ in a 22ml portion of diluted HCl. Cool the well stirred mixture thoroughly in ice-water and, when crystallization is complete, filter the bis-quinolinium tetrachlorozincate(II) with suction, wash with two 10ml portions of dilute HCl and drain well. Transfer the solid to a 250ml beaker; add a little water and then 10% NaOH solution until the initial precipitate of zinc hydroxide dissolves completely. Extract the quinoline with three 25ml portions of ether in a separatory funnel and dry the combined ethereal extracts with anhydrous calcium sulphate. Remove the ether by flash distillation using a 10ml flask and finally distill the residue from an air bath using an air condenser. Collect the quinoline at 236-238°C as a colorless liquid, the yield is 6.9g (50%). If the distillation preformed under reduced pressure, quinoline has Bp= 118-120°C / 20mmHg.

Experiment # 17

Preparation of 2,4,6-trimethyl quinoline

Introduction

1,3-Dicarbonyl compounds condense with primary aromatic amines to give anils, which on treatment with conc. H2SO4 yield quinolines (the Combes reaction). Interaction of acetylacetone with p-toluidine gives the anil, which is cyclized by conc. H_2SO4 to 2, 4,6-trimethyl quinoline.

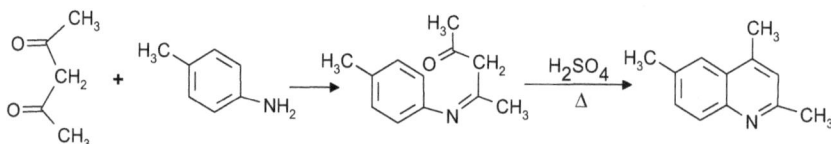

Procedures

1- Preparation of β- (p-toluidino)propenyl methyl ketone

In 250ml round-bottomed flask fitted with a Dean and Stark apparatus and reflux condenser, heat under reflux a solution of p-toluidine (10g) and acetylacetone (10g ; 10.2ml) in xylene (50ml). When the theoretical quantity of water (2ml) has been collected (after 2hrs), distill off the excess of solvent, then cool and triturate the residue with cold petroleum ether (B.p = 40-60°) until solidification occurs. Filter the mixture and wash the residue with cold petroleum ether (b.p. = 40 -60°; 2 x 30ml). β- (p-Toluidino) propenyl methyl ketone is obtained as pale - yellow plates, yield = 7.5 - 8g (39.5 - 42.5%).

2- Preparation of 2, 4, 6-trimethyl quinoline

To warm conc. H_2SO_4 (25ml) contained in a 250ml conical flask, add portion wise the prepared anil (4g). When addition is completed, heat the mixture on a water bath at 100° for 30min, then cool it and cautiously pour it into ice-water (200ml). Basify the resulting solution by adding solid Na_2CO_3 and filter off the precipitate product. 2, 4, 6-trimethyl quinoline dihydrate crystallizes (with difficulty) from 60% aqueous ethanol (8 ml/g) as colorless needles Mp= 62 - 63°. Yield = 3 - 3.2g (68.5-73%).

Experiment # 18

Preparation of 2, 3-dioxoindole (Isatin)

Introduction

Cyclization of isonitroso acetanilide under strong acid conditions (eg. polyphosphoric acid, PPA) yields 2, 3-Dioxoindole. Isonitrosoacetanilide is readily prepared from aniline, chloral hydrate and $NH_2OH. HCl$.

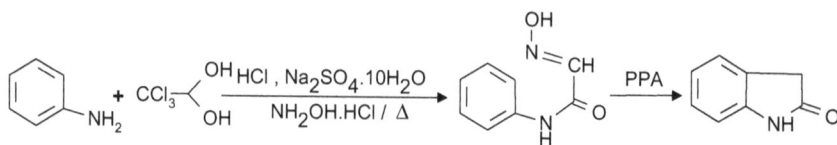

Procedures

1- Preparation of isonitrosoacetanilide

Dissolve freshly distilled aniline (4.65g; 4.5ml) in conc. HCl (4.3ml) and water, and to the solution add a solution of chloral hydrate (9g) and sodium sulphate decahydrate (130g) in water (130ml) . Next add a solution of hydroxylamine hydrochloride (11g) in water (150ml) and bring the resulting mixture to reflux temperature over a period of 15 min, then heat under reflux for 5 min. Allow the solution to cool and filter off the separated product. Wash the residue with cold water and dry at 100°. Isonitrosoacetanilide is obtained as buff plates, m.p. = 178-179°, sufficiently pure for direct use in the next stage (Yield =6.5-7.5g (79-91.5 %).

2- Preparation of 2, 3-dioxoindole

To isonitrosoacetanilide (1g) add polyphosphoric acid (20g) and stir the mixture at 56° for 6 hrs. Pour the resulting mixture into ice-water (100ml), stir vigorously, then filter off the precipitated product and wash with cold water. 2, 3-Dioxoindole crystallized from glacial acetic acid (8ml/g) as orange prisms (m.p.=197°. Yield = 0.6 - 0.7g (67-78%).

Experiment # 19

Preparation of 2-methyl-4-quinolone

Introduction

β-Keto ester condenses with aromatic amines under mild conditions to yield anils, which crystallizes on heating in an inert solvent to the corresponding 4-quinolones (the Conrad- Limpach synthesis). Interaction of ethyl acetoacetate with aniline at low temperature (< 100°) yields the anile that on heating in diphenyl ether at 200° cyclizes to give 2-methyl- 4-quinolone.

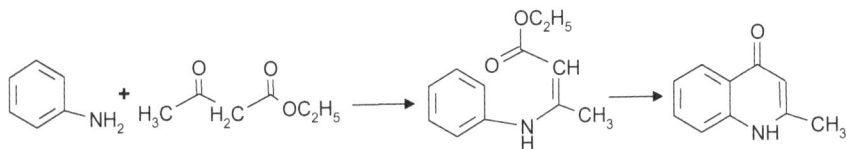

Procedures

1- Preparation of ethyl β-anilinocrotonate

In a 250ml round bottomed flask fitted with a Dean and Stark apparatus. In addition, reflux condenser, heat under reflux a solution of freshly distilled aniline (9.3g = 9.1ml) ethyl acetoacetate (13g=12.7ml) and glacial acetic acid (1ml) in benzene (100ml). When the theoretical quantity of water (2ml) has been collected (after 3hrs), evaporate off the solvent and transfer the crude product into a 25ml pear- shaped flask, and distill under water pumb vacuum. Ethyl β-anilinocrotonate is obtained as a paleyellow oil b.p.=162 -166°/12mm, yield=14.5-15g (70.5-73%).

2- Preparation of 2-methyl-4-quinolone

To boiling diphenyl ether (25ml), contained in a 250ml 2-necked, round-bottomed flask equipped with reflux condenser and dropped-funnel, add drop wise over 5min. ethyl β-anilinocrotonate (5.1g). Heat the solution under reflux for 30min, then cool and add petroleum ether (B.p= 60-80°C, 100ml). Stir the mixture until a white solid is formed, then filter and wash the residue with petroleum ether (b.p. = 60 - 80°C, 4 x 50ml). Final product crystallizes from water (20ml/g) as colorless prisms (m.p. =232 -234°), Yield = 3.1-3.5g (78 -88%)

Experiment # 20

Preparation of 1,2 ,3 ,4-tetrahydrocarbazole

Introduction

The phenyl hydrazone derivatives of acyclic ketones are readily cyclized to obtain indoles in the presence of Zinc chloride (Fischer indole synthesis). Borsch showed that the phenyl hydrazone of cyclohexanone undergoes a similar reaction to yield 2, 3,4-Tetrahydro carbazole. Glacial acetic acid is used in this case as the solvent for phenyl hydrazone formation and also as the cyclizing agent.

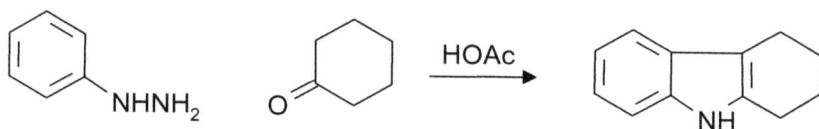

Preparation

1- To a solution of cyclohexanone (4.75g; 5ml) in glacial acetic acid (30ml) contained in a 100ml round-bottomed flask add (5.4g; 4.5ml phenyl hydrazine or 7g phenyl hydrazonium chloride/ 5g sodium acetate, anhydrous) .

2- Heat the mixture under reflux for 10min, then cool and filter off the brown solid.

3- Wash the product with cold water (4x50ml) and dry in a vacuum desiccator over conc. H_2SO_4.

Notes:

Tetrahydrocabazole must be dried in an oven, since it is readily oxidized into the hydroperoxide. 1,2,3,4-Tetrahydrocabazole crystallizes from 50% aqueous methanol (ca. 6ml/g) and remove the color by using decolorizing carbon to get white prisms product, m.p. =114-116°C, yield = 6-7g (70-80%)

Experiment # 21

Preparation of 4,5-diphentlimidazole(4,5-diphenylglyoxaline)

Introduction

In the Razisewski reaction, alfa-dicarbonyl compounds condense with NH_3 and aldehydes to form imidazoles. In this modified Razisewski reactions, binzil, hexamine and ammonium acetate react in acetic acid to yield 4, 5-diphenylimidazole.

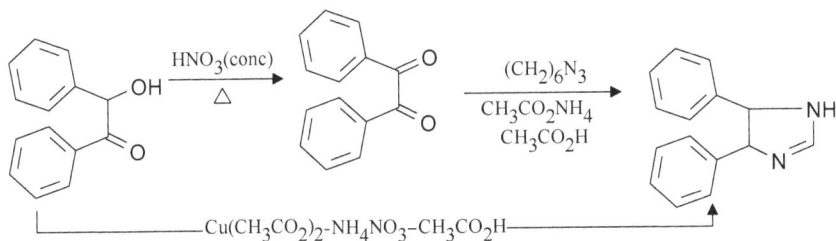

Preparation

Method (1)

1- Place (20g; 0.094mol) of crude benzoin and 100ml conc. HNO_3 in 250ml round-bottomed flask.

2- Heat on boiling water bath (in a fume cupboard) with occasional shaking until the evolution of nitrogen oxides are ceased (about 1.5hrs).

3- Pour the reaction mixture into 300-400ml of cold water containing in a beaker. Stir well until the oily crystallized completely as a yellow solids.

4- Filter the crude benzil at the pump, and wash it through with water to remove the HNO_3 acid. Recrystallize from ethanol or petroleum ether (2.5ml/g). The yield of pure benzil (19g) = 94-96%.

Method (2)

1- Place Copper (II) acetate (0.2g), ammonium nitrate (10g; 0.125mol), benzoin (21.2g; 0.1mol) and aqueous acetic acid solution (70ml; 80%; V/V) in 250ml round-bottomed flask fitted with refluxing condenser. Note when a solution is formed and a vigorous evolution of nitrogen oxides is observed.

2- Reflux for 90min, and then cool the reaction mixture.

3- - Seed the solution with crystal of benzil and allow to stand for one hour.

4- Filter at the pump and keep the mother liquor.

5- Wash well with water and dry (preferably in an oven at 60°).

The yield of benzil (19g, 90%), M.p= $94-95^O$. The M.p is unaffected by recrystallization from alcohol or carbon tetrachloride (2ml/g). Dilution of the mother liquor with the aqueous washing gives a further 1.0g of benzil .

Note:

1) For large-scale preparations use a three-necked flask equipped with two reflux condensers and a sealed mechanical stirrer.
2) Stirring or vigorous shaking also induces crystallization.
3) The mother-liquor should not be concentrated for an explosion may result.

Preparation of 4, 5-diphenylimidazole

Preparation

1- Place (2.1g) of benzyl, (0.26g) hexamine and (6g) ammonium nitrate in 100ml round-bottomed flask and add 50ml glacial acetic acid. Fit with refluxing condenser and reflux for one hour.
2- Add the resulting solution to 400ml water and remove any turbidity may develop by adding a small amount of carbon followed by stirring and filtration.
3- Precipitate the crude product by basifying the filtrate with conc. ammonium hydroxide (70ml) the cool the mixture.
4- Filter and wash the product with water. Purify the residue by dissolving it in hot pyridine (10ml), and carefully add hot water (0ml).
5- On cooling, 4,5-diphenylimidazole is obtained as fine elongated prism, m.p. = 233-234°, yield = 1.7 - 1.9g (77-86%).

Experiment # 22

Preparation of 5,5-diphenylhydartoin (Phenytoin)

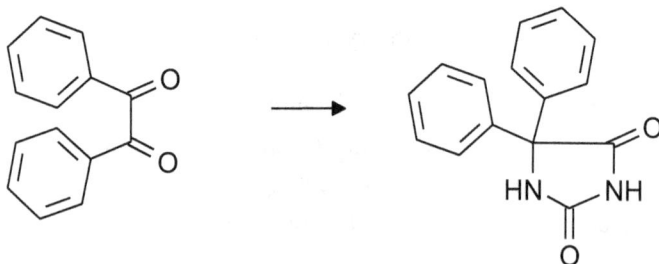

Preparation

1- Place benzyl (3.5g) and urea (2g) in 100ml round-bottomed and add NaOH (10ml of 30%) solution and alcohol (50ml).

2- Heat the reaction mixture using refluxing condenser and reflux for 1.5 hrs.

3- Cool to room temperature, then pour the contents into 150ml water and stir thoroughly.

4- Allow the mixture to stand for 15min and then filter.

5- Collect the filtrate into a clean beaker and make it strongly acidic with conc. HCl while stirring, then cool in an ice bath. Filter the product and wash with cold water and dry.

6- Purification of crude 5, 5-diphenylhydatoin can be done for recrystallization from methanol or spirit to obtain 1.4g (44%) yield with m.p. = 297-298°·

Questions

1. List all possible impurities in the cyclohexene sample prepared.
2. Calculate the theoretical amount of water that could be obtained from this reaction.
3. Write the equations for the reaction involved in this preparation and suggest a reasonable mechanism.
4. What alkene would be produced on dehydration of each of the following alcohol
   a) 1-Methylcyclohexanol   b) 2,2-Dimethylcyclohexanol
   c) 2-Methylcyclohexanol   d) 1,2-Cyclohexanediol
   e) 4-Methylcyclohexanol
5. Explain why the synthetic process used here permits us to carry the reaction essentially to completion.
6. Show how you would alter the conditions so that hydration of cyclohexene will be predominant reaction.
7. In the work-up procedure for cyclohexene, why is salt added before the layers are neutralized and separated.
8. What is the purpose of adding the sodium carbonate solution? Give an equation.
9. Compare and interpret the infrared spectra of cyclohexanol.
10. Write the equation for the hydration of 2-methyl-2-butene and 2-pentene. What is the function of H_2SO_4?
11. List the various isomeric amyl alcohols in order of decreasing ease of dehydration.
12. Can adipic acid be made conveniently from cyclohexanene as well as from cyclohexanone? Account for this reason.
13. Which of the following compounds will give haloform reaction product?
    (a) $C_6H_5Cl_3$ (d) $C_6H_5CH_2COCH_3$
    (b) $C_6H_5CHOHCH_3$ (e) $C_6H_5CH_2CHOHCH_3$
    (c) $C_6H_5COCH_3$ (f) $C_6H_5CH_2CH_2CH_3$
14. (a) What volume (in ml) of 1.00N base solution will be required for the neutralization (its equivalent weight as determined by neutralization with a standardized solution of base) of benzoic acid.
    (b) Calculate the neutralization equivalent (its equivalent weight as determined by neutralization with a standardized solution of base) of benzoic acid.
15. What is the neutralization equivalent of an acid?
    (a) If 1g require 16.67ml of 1.00N base

(b) If 1g of acid require 12.05ml of normal base.

16. Calculate the pH of a solution that is 0.10M in acetic acid and 0.20M in sodium acetate. The ionization constant of acetic acid is 1.8×10^{-5}.

17. A liquid of compound A has the formula $C_5H_{10}O$. Reaction of bromine in alkaline solution converts this compound to bromoform and sodium n-butyrate. What is molecular structure would you assign to A?

18. In the reaction of bromine in alkaline medium transform a compound B into bromoform and sodium salt of an acid. The purified acid has a neutralization equivalent of 88. Suggest two possible structures for B.

19. A liquid of compound C forms an oxime product that contains 13.88% of nitrogen. Compound C dose not reduce Fehling's solution nor give the haloform reaction. What is the molecular structure would you assign to C?MRAAM

20. Write the equations for the reactions used in the synthesis of cyclopentanone.

21. Crude cyclopentanone is washed with sodium carbonate solution. Explain why?

22. Propose another method for synthesis of cyclopentanone and write an equation for the reaction used.

23. What are the products obtained from dry distillation of the following acids: Malonic, Glutaric, Succinic. In addition, Pimelic acid.

24. Why the final washing in the preparation of cyclopentanone is done with sodium chloride solution rather than pure water.

25. What experimental method would you recommend for the preparation of n-octyl bromide and t-Butyl bromide?

26. Explain why the crude product is a part to certain definite organic impurities?

27. Sulphuric acid was used to remove unreacted alcohol from the crude alkyl halide. Explain how it removes the alcohol. Write the equation?

28. Di-butyl ether and 1-butene can be formed as by-products in the synthesis of n-octyl bromide. Explain how they are removed by sulphuric acid. Give the reactions.

29. Look up the density of n-butyl chloride. Assume that this alkyl halide was prepared instead of the bromide. Decide whether the alkyl halide would appear as the upper or the lower in the separatory funnel at each stage of the isolation; after the reflux,

after the distillation, after the addition of water to the distillate, after the washing with sulphuric acid and then after the washing with NaOH.

30. Why most the crude alkyl halide be dried carefully with calcium chloride before the final distillation?

31. What is the purpose of the conc. H_2SO_4 acid used in the acylation reaction?

32. Give a possible structure of the polymeric by-product obtained in the acylation reactions.

33. Why is the polymeric by-product not soluble in Na_2CO_3 solution, while salicylic acid itself is soluble?

34. If one were to use 5.0g of salicylic acid and excess acetic anhydride in the synthesis of aspirin, what would be the theoretical yield of acetylsalicylic acid in mol and in grams?

35. When aspirin is heated in boiling water, it decomposes. The resulting solution gives a positive Iron(III) chloride test. Why is this test positive? Give the equations for the reaction.

36. HCl is about as strong a mineral acid as H_2SO_4, why would it not be a satisfactory catalyst in the reaction of aspirin synthesis?

37. How do you account for the smell of vinegar when an old bottle of aspirin is opened?

38. The equilibrium constant for the formation of ethyl acetate from ethanol and acetic acid at 25° is 3.77. Calculate the percentage of ester present at equilibrium with the molar ratio of reactants used in the experiment.

39. Write a detailed mechanism for the Fischer esterification reaction.

40. State the principle of microscopic reversibility in your own words. How does this principle relate to today's experiment?

41. How is the rate of Fischer esterification affected by alkyl substituent in α- and β - positions of the carboxylic acid?

42. Write series of equations showing phenol may be used for the production of a) cyclohexanone oxime, b) cyclohexanone and c) diethyl adipate.

43. In the reaction of diethyl adipate with ammonia, would you expect to obtain adipamide $(CH_2CH_2CONH_2)_2$ or the cyclic imide?

44. At what relative rates would you expect the following esterification proceeds?

a) $CH_3COOH + n -CH_4OH_9$

b) $C_2H_5CH(CH_3)COOH + n-C_4H_9OH$

c) $C_2H_5CH(CH_3)COOH + sec\text{-}C_4H_9OH$

d) $C_2H_5CH(CH_3)COOH + tert\text{-}C_4H_9OH$

e) $(CH_3)_3COOH + tert\text{-}C_4H_9OH$

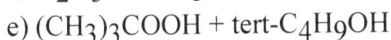

45. Why the potassium salts of fatty acids yield soft soaps?
46. Why is the soap derived from coconut oil so soluble?
47. Why does adding a salt solution cause soap to precipitate?
48. Why do you suppose a mixture of ethanol and water instead of a simply water itself is used for saponification?
49. Sodium acetate and sodium propionate are poor soap. Why?
50. Write down the reactions used in the synthesis of benzalacetophenone and write a reasonable mechanism for the formation processes.
51. In the reaction carried out for the synthesis of benzalacetophenone, what may be two significant side reaction?
52. What problems may be encountered if:
    a) Benzaldehyde is added to KOH solution and left for one hour and acetophenone is added after this period.
    b) Acetophenone is added to KOH solution and left for one hour and benzaldehyde is added after this period.
53. Write a mechanism for the aldol condensation of one mole of benzaldehyde with one mole of acetone.
54. Write the mechanism for the base-catalyzed dehydration of the product from the reaction in Question 53 to give benzalacetone.
55. Why is it important to maintain equivalent proportions product reagent in the above reaction?
56. What side products do you expect in the above reaction? How are they removed?
57. What do the melting points of the crude and recrystallized products tell you about purity in the above?
58. Describe two methods of the reduction of Nitrobenzene to Aniline.
59. Why before the extraction with ether in the synthesis of aniline from nitrobenzene, the reaction mixture is made alkaline with NaOH solution?
60. Write the advantages of steam distillation.
61. Explain why the acetamido group is an ortho, para directing?
62. Outline the acid catalyzed hydrolysis of p-nitroacetanilde to yield p-nitroaniline.
63. o-Nitroaniline is more soluble in ethanol than p-nitroaniline.

64. N-methylbenzamide, an isomer of acetanilide, when allowed to react with HNO_3-H_2SO_4 mixture gives a different product from what is obtained from acetanilide.
65. Write the mechanism for the reaction of p-toluidine with acetic anhydride. Why sodium acetate is added for this reaction?
66. In the oxidation step p-toluidine, if excess of permanganate remains after the reaction period, a small amount of ethanol is added to discharge the purple color. Write the chemical equation that describes the reaction of $KMnO_4$ with ethanol.
67. Write the reaction mechanism for the acid-catalyzed hydrolysis of p-acetamido benzoic acid to form p-aminobenzoic acid.
68. What is the chemical structure of the precipitate that forms after the sulphuric acid has been added in the hydrolysis of p-acetamido benzoic acid?
69. What kind of gases were evolved during the experiment in the above process?
70. What is the structure of the inorganic solids that formed the hydrolysis of p-acetamido benzoic acid after the neutralization of the aqueous mixture which has been decanted from the solid? The benzocaine did not precipitate during the neutralization. Why not?
71. HNO_3 acid is generated by the action of H_2SO_4 acid on $NaNO_3$. Nitrous acid is prepared by the action of HCl on sodium nitrite. Why nitrous acid is prepared in situ, rather than obtained from the reagent shelf?
72. What by-product would be obtained in high yield if the diazotization of p-toluidine were carried out at 30°C instead of 0-5°C?
73. Write a balanced equation for the reduction of nitrobenzene to aniline using Tin and HCl as reductants. Tin will be in the (+) oxidation state at the end of the reaction.
74. Write the balanced equation and mechanism for the hydrolysis of methyl benzoate with aqueous acid.
75. Ordinarily amines are isolated from Tin/HCl reduction by adding sufficient NaOH to convert tin salt into H_2O soluble sodium stannite and then isolating the product. Why does this procedure not work in the reduction of nitrobenzene to aniline experiment?

Chapter IV

Quantitative Analysis of Chemical Compounds

IV.1 Volumetric analysis

IV.1.1 Introduction

Analytical chemistry is a branch of chemistry that deals with the use of tools for chemical analysis. This branch characterized by a good accuracy and rapidity, in which, the ratios of the titrated substance components can be calculated, depends on the form of the reactants that should be solutions. This can be carried out using conical flasks and burettes and other tools to facilitate this kind of analysis. One the solutions (which present in the burette) can be added to the other one (in the conical flask) in which the chemical reaction can be completed and can be indicated by using an indicator (few drops can be added) to establish the end point of the reaction. This process is known as titration

IV.1.2 Primary and secondary substances

The primary substances are those substances of high molecular weight, high purity and non-hygroscopic, for example; sulphonic acid and sodium carbonate (used in neutralization titration), silver nitrate (used in precipitation titration), sodium oxalate, potassium dichromate and potassium iodate (used in oxidation-reduction titration). The secondary substances are those substances of low molecular weight, low purity and hygroscopic, for instance; H_2SO_4 (used in titration with sodium carbonate).

IV.1.3 Primary and secondary solutions

The primary solution was prepared by weighing known concentration of primary standard substance in known volume of solvent. Meanwhile, the secondary solution was prepared by dissolving approximate weight of the substance in distilled water or other solvent then titrated by a primary solution to justify the concentration.

IV.1.4 Classification of titrations

The titration process is classified into two kinds:

I- Process with no electron transfer reactions

1- Acid-base titration

This process depends upon the reaction between hydrogen ion of the acid and the hydroxyl ion of the base which forms a salt and water, and this reaction is known as neutralization titration.

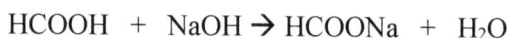

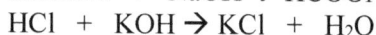

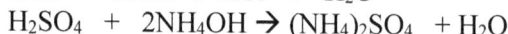

$$HCOOH + NaOH \rightarrow HCOONa + H_2O$$
$$HCl + KOH \rightarrow KCl + H_2O$$
$$H_2SO_4 + 2NH_4OH \rightarrow (NH_4)_2SO_4 + H_2O$$

The neutralization titration is divided into four types, due to the difference in the strength of the acids and bases and they can be summarized as strong acid - strong base, Strong acid- weak base, Strong base- weak acid and Weak acid - weak base titrations

2- Precipitation titration

This process involves two solutions. One of them pptd the other forming new products which include substance insoluble in water and the substance as solution, to illustrate this titration, the reaction between silver nitrate solution and potassium chloride solution forms a product of silver chloride as white ppt (insoluble in water) and potassium nitrate as a solution (soluble in water) according to the following chemical reaction:

$$Ag NO_3 + KCl \rightarrow AgCl\downarrow + KNO_3$$

At the end point, the solution of excess Cl^- is changed to a solution containing an excess of Ag^+, and the potassium chromate is used as indicator to establish the end point of the reaction.

3- Complexometric titration

This method is used for example to estimate the hardness of water by certain specific chemicals such as ethylenediaminetetraacetic acid (Known as EDTA) as complex agent for several metal ions, in which there will be formed with very stabilized coordination compounds with transition and non-transition metal ions. Each molecule of EDTA bonded to metal ion through a coordinate bond, because it has six atoms which can be involved in the coordination and considered to be hexadentate ligand (two nitrogen atoms and four oxygen atoms). This kind of analysis depends on the stability of the complexes. This is referring to what called spectrochemical Series:

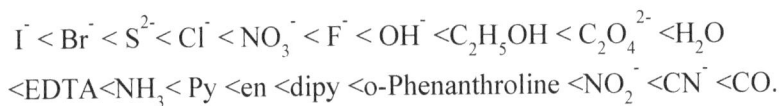

$$I^- < Br^- < S^{2-} < Cl^- < NO_3^- < F^- < OH^- < C_2H_5OH < C_2O_4^{2-} < H_2O$$

$$< EDTA < NH_3 < Py < en < dipy < o\text{-Phenanthroline} < NO_2^- < CN^- < CO.$$

The EDTA earns the stability of its complexes from its special chemical structure which is explained below, and which can be used as acid (M= H) or as sodium salt (M = Na) in the titration, the chemical structure of the salt is:

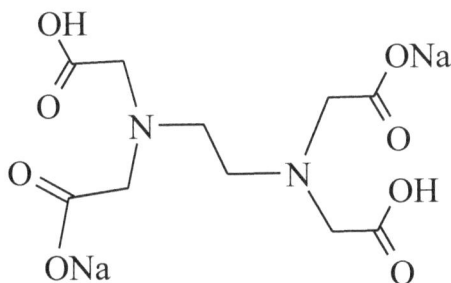

II- Titration process with electron transfer (Oxidation-reduction reactions)

In general, the oxidation is the reaction in which an element or compound combined with oxygen, and the reduction; is a reaction in which oxygen is removed partially or completely from a compound. These reactions are a type of chemical analysis, which involves the transfer of electrons during the chemical reaction in which they should be classified separately as oxidation- reduction reactions. The other definitions; is that the oxidation process is the removal of electrons from one reactant, and reduction process is the gain of electrons by anther reactant in the same process. The total number of removed electrons is the same as the gained electrons in these chemical reactions. The substance that gained electrons are called oxidizing agents (oxidants, for example; $KMnO_4$, $K_2Cr_2O_7$), meanwhile, the substances that losses electrons called reducing agents (reductants, for example; H_2CO_4, $FeSO_4$. The following chemical reactions illustrate the oxidation and reduction processes. The reaction between dichromate and iron (II) ions in the presence of sulphuric acid as media produces chromium (III) and iron (III) ions and water can be written as:

$$Cr_2O_7^{2-} + Fe^{2-} \rightarrow 2\,Cr^{3+} + Fe^{3+} + H_2O \qquad (1)$$

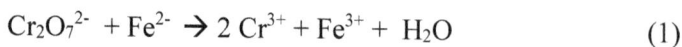

(i) Half reaction:

$$Cr^{6+} + 3e^- \rightarrow 2Cr^{3+} \qquad \text{Reduction process}$$

(ii) Half reaction:

$$Fe^{2+} \rightarrow Fe^{3+} + e^- \qquad \text{Oxidation process}$$

To balance the equation (1); multiply the equations; (i) by 2 and (ii) by 6, then add them together you will get:

$$2Cr^{6+} + 6e^- + 6Fe^{2+} \rightarrow 2Cr^{3+} + 6Fe^{3+} + 6e^-$$

Then the final oxidation reduction reaction equation becomes:

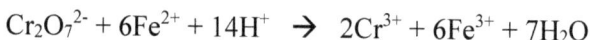

$$Cr_2O_7^{2-} + 6Fe^{2+} + 14H^+ \rightarrow 2Cr^{3+} + 6Fe^{3+} + 7H_2O$$

IV.1.5 The masses (weights)

I- The mole

The most important uses of atomic masses (or weights) is to find the relative masses of compounds or substances taken part in chemical changes or reactions. The mole can be defined as the amount or a formula weight of a substance expressed in grams.

II- Formula and molecular masses (or weights)

Formula weight is simply defined as the sum of the masses of the atoms in the empirical formula, however, the mass of the molecule is the sum of the masses of the atoms in the molecule. For example,

Molecular weight of H_2 is $2(1.008) = 2.016$ g/mol
Molecular weight of H_2O is $2(1.008) + (16.00) = 18.02$ g/mol
Molecular weight of H_2SO_4 is $2(1.008) + (32.06) + 4(16.00) = 98$ g/mol
Molecular weight of H_3PO_4 is $3(1.008) + (30.97) + 4(16.00) = 97$ g/mol

III- Equivalent weights

The equivalent weight is simply the weight of substances that are equivalent to one another in chemical reaction.

1- Equivalent weight of acids

The molecular weight **(M. Wt.)** of substance is the molecular weight **(M. Wt.)** of an acid divided by the number of the hydrogen ion replaced.

Eq. Wt. = Molecular weight/ Basicity
            (Basicity is the hydrogen ions replaced)
For example:
   Equivalent weight of HCl   is 36.5/1 = 36.5 g/mol
   Equivalent weight of H_2SO_4 is 98/2   = 49 g/mol
   Equivalent weight of H_3PO_4 is 97/3   =32.33 g/mol

2- Equivalent weight of bases

It is the molecular weight of the base divided by the number of the hydroxide ions replaced.

Eq. Wt. = Molecular weight/ Acidity
            (Acidity is the hydroxide ions replaced)

For example:
   Equivalent weight of KOH   is 56/1   = 56 g/mol
   Equivalent weight of $Ca(OH)_2$ is 74/2 = 37 g/mol
   Equivalent weight of $Al(OH)_3$ is 78/3 = 26 g/mol

3- Equivalent weight of a salts

It is the molecular weight of the salt divided by the number of the metal ions multiplied by their oxidation state

Eq. Wt. = Molecular weight/ n
            (n is the number of metal ions x their oxidation state)
For example:
   Equivalent weight of NaCl   is 38.5/1 = 38.5 g/mol
   Equivalent weight of Na_2SO_4 is 142/2 = 71 g/mol
   Equivalent weight of $(Al)_2(SO_4)_3$ is 342/6 = 57 g/mol

4- Equivalent weight of oxidation-reduction reactions

It is the molecular weight of the oxidizing agent divided by the number of the gained electrons and the weight of the reducing agent divided by the number of lost electrons.

Eq. Wt. = Molecular weight/ n
        (n is the number of gained or lost electrons)

For example:
  Equivalent weight of $K_2Cr_2O_7$ is 294/6 = 49 g/mol
  Equivalent weight of $FeSO_4.7H_2O$ is 278/1 = 278 g/mol
  Equivalent weight of $KMnO_4$ is 158/5 = 31.6 g/mol

5- General rules of equivalent weight of the substances

Acid–base titrations

Equivalent weight is the formula weight divided by the number of hydrogen atoms in the acid, whereas in the base is the number the required hydrogen atoms needed to neutralize each base molecule.

(Eq. Wt.)$_{Acid - base\ titrations}$ =   **Formula wt.**
                                        Number of H^+

Oxidation-reduction reaction

Equivalent weight is the formula weight divided by the number of hydrogen atoms in the acid, whereas in the base is the number the electrons lost or gained in the chemical reactions.

(Eq. Wt.)$_{Redox}$ =   **Formula wt.**
                         Number of e^-

Precipitation reactions

Equivalent weight of the metal ion is the formula weight of the metal ion divided by the number of the charge on the ion.

(Eq. Wt.)$_{Precipitation\ reaction}$ =   **Formula wt.**
                                           Ion charge

IV.1.6 Preparation of standard solutions

I- Molar solutions (M = M. Wt$_{solute}$/Liter $_{solvent}$):

This kind of concentrations can can prepared by dissolving one gram molecular weight of the substance in a suitable solvent and complete up to a liter to get 1 molar (1M). For example, a molar solution of HCl; contains 36.5g of the acid dissolved in one liter distilled water. To prepare 0.25M of this acid; 9.2g dissolved in one liter of distilled water.

$$M = \frac{Grams\ x\ 1000}{Molecular\ weight\ x\ required\ volume}.$$

Therefore,

The molarity is the concentration of a solute in solution, expressed in moles per liter (M = mol/L)

To illustrate this, we can follow the following example.

Example: Calculate the Molarity of oxalic acid (M. Wt. = 126.00 g/mol) solution when 4.5g of an acid was dissolved in 350 ml of distilled water.

Solution To solve this problem, we have to use the following.

The formula of the oxalic acid is $H_2C_2O_4.2H_2O$
The formula weight of the acid is 126.00g/mol

Therefore, the Molarity according to the previous relation =

$$M = \frac{4.5\ x\ 1000}{126\ x\ 350} = 0.102$$

II- Normal solutions (N = Eq. Wt$_{solute}$/Liter$_{solvent}$):

Normal solutions can be prepared by dissolving one gram equivalent weight of a substance in suitable solvent then the volume completed into one liter (using measuring flask). For instance, 1Normal (1N; called one normal solution) of sulfuric acid contains 49g of the acid dissolved in one liter distilled water. To prepare 0.1N of this acid; 4.9g of the acid dissolved in one liter of distilled water.

$$N = \frac{Grams \times 1000}{Equivalent\ weight \times required\ volume}.$$

Therefore,

The Normality is the concentration of a solute in solution, expressed in equivalent weight per liter (N = Eq. Wt /L)

To illustrate this, we can follow the following example.

Example: Calculate the normality of an acid (M. Wt.= 49g/mol) solution when 9.8g of an acid was dissolved in 300ml of distilled water.

Solution: To solve this problem, we have to use the following relation:

$$N = \frac{9.8 \times 1000}{49 \times 300} = 0.67$$

III- Relationship between Molarity (M) and Normality (N) In general the relation between the molar weight and the equivalent weight can be defined. For monobasic acids the molarity equals the normality, However in dibasic acids the relation can be expressed as; **Normality = 2 Molarity**

To elucidate this relation, we can follow the example:

Example: Calculate the normality of 0.15M of sulfuric acid (dibasic) solution.

Solution

By applying the relation of (N = 2M), then the normality of the acid:

$$N = 2 \times 0.15 = 0.30$$

IV.1.7 Titration methods

There are two methods:

I- Direct method

The goal of this titration is the addition of the titrant to the titrated solution (analyzed substance) until the equivalent point appeared.

II- Indirect method

Sometimes the rate of the chemical reactions is low for titration to be preformed, directly by the addition of the titrant. Therefore, this process, the standard solution is added in excess to the solution which is required to be analyzed and then the volume of the excess standard solution was determined by titration and the normality of the analyzed solution and then be calculated.

III- Indicators

Those are defined as weak organic compounds whose colors changed according to the medium; the media means acidic, basic, or neutral media. The indicators were used in all titration processes to identify the equivalent points (end points). There are varieties of indicators used in acid - base titration such as; methyl orange, methyl red, phenolphthalein, methyl yellow and litmus paper...etc. Ferrum alum, fluorescen and potassium chromate indicators were used in the precipitation titration. Whereas, Eriochrome black-T, Murexide and Pyrochatechol were added in complexometric titration. For oxidation reduction titration, a permanganate solution considered to be self indicator. Table-IV-1 contain most of the used indicators in analysis with the concentration, as well as the pH ranges.

Table IV-1 pH range of the indicators

Indicator	conc.	Solvent	pH $_{range}$	Media/ Metal ions
Methyl Orange		20% alcohol	3 - 4.5	Red (acidic), Orange (basic)
Methyl Yellow			2 - 3	Red (acidic), Yellow (basic)
Methyl red		50% alcohol	4.4 - 6.6	Red (acidic), Yellow (basic)
Phenolphthalein	1.0%	60% alcohol	8.3 - 10	Colorless (acidic), Red (basic)
Litmus paper	4%	water	5 - 8	Red (acidic), Blue (basic)
Eriochrome black-T			6.3 - 11.6	Ca^{2+}, Mg^{2+}
Murexide			7	Ca^{2+}, Co^{2+}, Ni^{2+}, Cu^{2+}
Pyrochatechol			2 - 6	Bi^{3+}
Yellow Alizarin	0.01%	70%	10.0 -	Yellow (acidic),

G		alcohol	12.1	Red (basic)
Bromophenol blue	0.04%	20% alcohol	2.8 – 4.6	Yellow (acidic), Violet (basic)
Bromocresol green	0.04%	20% alcohol	3.6 – 5.2	Yellow (acidic), Blue (basic)
Thymol blue	0.04%	20% alcohol	1.2 - 2.8	Red (acidic), Yellow (basic)
Phenol red	0.02%	95% alcohol	6.8 – 8.4	Yellow (acidic), Violet (basic)
Cresol red	0.02%	20% alcohol	0.2 – 1.8	Yellow (acidic), Red (basic)
Congo red	0.5%	water	3.0 – 5.0	Blue (acidic), Red (basic)
Neutral red	0.01%	50% alcohol	6.8 – 8.0	Red (acidic), Orange (basic)

IV.1.8 Experimental applications

We offer in this section several experiments to represent most types of the titration.

I- Strong acid with strong base titration

This kind of titration aims at the calculation of the normality (N) and the amount of the hydrochloric acid by using a primary standard solution of sodium hydroxide (0.1N).

1- Principle of the experiment

In the presence of phenolphthalein indicator, HCl reacts with NaOH forming NaCl and H_2O according to the following reaction. The color of the indicator is pink in basic media and becomes colorless at the end point. HCl + NaOH → NaCl + H_2O

2- Required chemicals and apparatus

Hydrochloric acid, sodium hydroxide solutions, phenolphthalein, burettes, pipettes, beakers, conical flasks, funnels and stands.

3- Procedures

1- Wash the burette with tap water and then with distilled water.

2- Rinse it out with small portion of the proposed HCl solution.

3- Fill the burette almost to the top with HCl solution.

4- Transfer 10ml of NaOH (0.1N) to a conical flask, and then add 2-3 drops of indicator.

5- Titrate HCl solution by NaOH solution (let the HCl dropdown from the burette to the conical flask which contains NaOH solution) with stirring until the end point appeared.

6- Repeat this experiment three times and take the average of the readings, then calculate the normality and the amount of HCl.

4- Calculations

Milliequivalents of HCl = Milliequivalents of NaOH ($N_a V_a = N_b V_b$).

Suppose the volume obtained from the burette is 8ml

$$N_{HCl} = \frac{0.1 \times 10}{8} = 0.125N$$

Amount of HCl = 0.125 x 36.5 = 4.56 g/l

II- Strong acid with weak base

This kind of titration aims for the calculation of the normality (N) and the amount of the H_2SO_4 by using a primary standard solution of sodium carbonate (0.1N).

1- Principle of the experiment

In the presence of methyl orange indicator, the H_2SO_4 reacts with Na_2CO_3 forming a salt (Na_2SO_4), H_2O and CO_2, according to the following chemical reaction. The color of the indicator is yellow in basic media and becomes red at the end point. $H_2SO_4 + Na_2CO_3 \rightarrow Na_2SO_4 + CO_2 \uparrow + H_2O$

2- Required chemicals and apparatus

H_2SO_4, Na_2CO_3, methyl orange, burettes, pipettes, beakers, conical flasks, funnels and stands.

3- Procedures

1- Wash the burette with tap water and then with distilled water.

2- Rinse it out with small portion of the proposed H_2SO_4 solution.

3- Fill the burette almost to the top with H_2SO_4 solution.

4- Transfer 10ml of Na_2CO_3 (0.1N) to a conical flask, then add 2-3 drops of indicator.

5- Titrate Na_2CO_3 solution by H_2SO_4 solution (let the H_2SO_4 solution drop down from the burette to the conical flask which contains Na_2CO_3 solution) with continuous stirring until the end point reached.

6- Repeat the titration three times and take the average of the readings, then calculate the normality and the amount of H_2SO_4.

4- Calculations

Milliequivalents of H_2SO_4 = Milliequivalents of Na_2CO_3

$$N_a V_a \ = \ N_b V_b$$

Suppose the volume obtained from the burette is 8.5ml

$$N(H_2SO_4) = \ \frac{0.1 \times 10}{8.5} = 0.117N$$

Amount of H_2SO_4 = 0.117 X 49 = 5.73 g/l

IV.1.9 Precipitation titration

This kind of titration aims to find the purity percentage of HCl in a sample of 250ml.

1- Principle of the experiment

In the reaction of silver nitrate solution with HCl solution, a white ppt of silver chloride is formed according to this reaction:

$$AgNO_3 + HCl \rightarrow AgCl\downarrow + HNO_3$$

After filtration process, discard the ppt and keep the filtrate (silver nitrate) which is unreacted with the acid to react with potassium thiocyanate of known concentration.

2- Required chemicals and apparatus

$AgNO_3$ (0.1N), HCl, Ferrum alum, burettes, pipettes, beakers, conical flasks, funnels and stands.

3- Procedures

1- Wash the burette with tap water and then with distilled water.
2- Rinse it out with small portion of the proposed $AgNO_3$ solution.
3- In a conical flask, weigh 250ml of concentrated HCl and complete the volume by distilled water up to the given mark.
4- Shake the solution and transfer 10ml to a conical flask followed by the addition of 5ml of 6N HNO_3 and 50ml of $AgNO_3$ solution (0.1N).
5- A white ppt of AgCl will appear, heat the contents of the conical flask, then filter and wash the obtained ppt by dilute HNO_3 solution.
6- Fill the burette almost to the top with potassium thiocyanate KSCN solution.
7- Add 1ml of the used indicator to the filtrate and then titrate by KSCN (0.1N) until the color of the solution becomes a pale brown color.
8- Repeat this experiment three times and take the average of the readings, then calculate the purity percentage of the acid.

4- Calculations

Suppose the weight of 2ml of conc. HCl is 2.2g and 30ml of KSCN solution (0.1N) was consumed to balance the excess of unreacted $AgNO_3$ with HCl. The volume of the unreacted $AgNO_3$ solution can be calculated as:

Milliequivalents of KSCN = Milliequivalents of $AgNO_3$

$$NV (AgNO_3) = NV(HCl)$$

$$V_{(AgNO3)} = \frac{0.1 \times 40}{0.1} = 40ml \qquad N_{(AgNO3)} = \frac{(N)HCl \times (V-40)HCl}{0.1}$$

Then, the normality of HCl   = $(0.1 \times 10) \ 10 = 0.1N$
The weight of HCl in 250ml   = $0.1 \times 36.5 \times 0.25 = 0.9125g$
The purity percentage of the HCl = $(0.9125/2.5) \times 100 = 36.5 \%$.

IV.1.10 Complexometric titration

This kind of titration aims at estimating the hardness of water.

1- Principle of the experiment

The estimation of the cations as carbonate were done by titrating against ethylenediaminetetraacetic acid (EDTA) using Eriochrome black-T indicator (EBT) and ammoniacal buffer solution (pH = 10). The EDTA molecule is polydentate ligand (Hexadentate) that means it has six sites of coordination which form bonds with metal ion through two amino and four carboxyl groups. The EDTA has chemical structure as shown previously and the indicator has chemical structure as shown below:

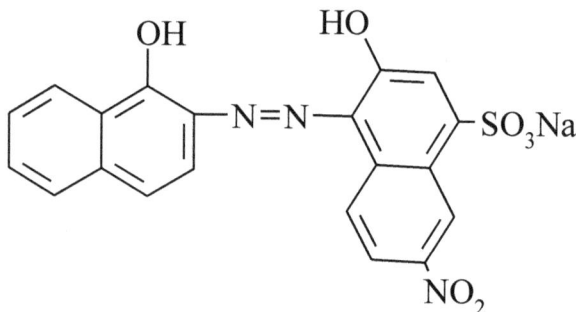

By considering EDTA, we can estimate the amount of calcium and magnesium ions as carbonate in part per million units (ppm). The end point of the reaction is blue color. The following two reactions occur in complexometric titration:

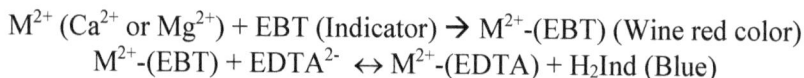

M^{2+} (Ca^{2+} or Mg^{2+}) + EBT (Indicator) $\rightarrow$ M^{2+}-(EBT) (Wine red color)
M^{2+}-(EBT) + EDTA^{2-} $\leftrightarrow$ M^{2+}-(EDTA) + H_2Ind (Blue)

2- Required chemicals and apparatus

Tap and distilled water, 0.01M EDTA, Eriochrome black-T(EBT), Ammoniacal buffer solution (pH=10), Burettes, Pipettes, Conical flasks, Beakers, measuring cylinders and stands.

3- Procedures

1- By pipette, transfer 10ml of tap water into 250ml conical flask, and then add about five ml of buffer solution and small amount of the EBT indicator to the tap water.
2- Titrate the mixture with 0.01M EDTA from the burette until the color changes from wine red to blue.
3- Repeat this experiment three times, until three concordant reading or take the average of the readings, then calculate the amount of calcium carbonate in ppm.

4- Calculations

Suppose the average reading of the burette is 2.5ml. The amount of $CaCO_3$ is calculated as: milliequivalents of EDTA = milliequivalents of $CaCO_3$

$$MV = M_1V_1$$

$$0.01 \times 2.5 = M_1 \times 10$$

$$M_1 = (0.01 \times 2.5) / 10 = 0.0025$$

Amount of $CaCO_3$ = 0.0025 x 100($CaCO_3$ M. Wt.) x 1000 = 250 ppm.

IV.1.11 Oxidation-reduction titration

This kind of titrations aim for the calculation of:

The normality of and the amount of $KMnO_4$ (Oxidizing agent) by titrating with 0.1N of Mohr salt; ammonium ferrous sulphate (Reducing agent).

The normality of and the amount of $K_2Cr_2O_7$ (Oxidizing agent) by titrating with 0.1N of Oxalic acid.

Experiment-1): Principle of the experiments

In the presence of H_2SO_4, $KMnO_4$ reacts with Iron (II) producing manganese (II), Iron (III) and water as shown in the following equations:

The ionic reactions are:

Reduction process $Mn^{7+} + 5e \rightarrow Mn^{2+}$
Oxidation process $\quad\quad Fe^{2+} \rightarrow Fe^{3+} + e$.

The overall reaction:

$$MnO_4^- + 5Fe^{2+} + 8H^+ \rightarrow Mn^{2+} + 5Fe^{3+} + 4H_2O.$$

1- Required chemicals and apparatus

$KMnO_4$, ammonium ferrous sulphate. hexahydrate; $(NH_4)_2Fe(SO_4)_2.6H_2O)$, H_2SO_4 , distilled water , Burettes, Pipettes, Conical flasks, Beakers, measuring cylinders and stands.

2- Procedures

1- By pipette, transfer 10ml of iron (II) solution into 250ml conical flask.
2- Titrate the $KMnO_4$ solution with 0.1N of iron (II) solution until the color of the solution becomes pink.
3- Repeat this experiment three times, until three concordant reading or take the average of the readings then calculate the amount of $KMnO_4$ in gram per liter.

3- Calculation

Suppose the required volume of $KMnO_4$ consumed for titration process is 7.5ml; Milliequivalents of Fe^{2+} = Milliequivalents of $KMnO_4$.

$$NV = N_1V_1$$

$$V_1(KMnO_4) = (0.1 \times 10) / 7.5 = 0.133$$

The amount of $KMnO_4$ = 0.133 x 31.6 = 4.20 g/l

Experiment-2): Principle of the experiments

In the presence of H_2SO_4, potassium dichromate reacts with oxalic acid producing chromium (III), carbon dioxide and water as shown in the following equation:

The ionic reactions are:

Reduction process $Cr^{6+} + 6e \rightarrow 2Cr^{3+}$
Oxidation process $\quad\quad C_2O_4^{2-} \rightarrow 2CO_2 +2e$.

The overall reaction:

$$2Cr_2O_7^{2-} + 6C_2O_4^{2-} + 28H^+ \rightarrow 4Cr^{3+} + 12CO_2 + 14H_2O.$$

1- Required chemicals and apparatus

Potassium dichromate, oxalic acid, mixture of H_2SO_4, diphenylamine, distilled water. Burettes, Pipettes, Conical flasks, Beakers, measuring cylinders and stands.

2- Procedures

1- By pipette, transfer 10ml of oxalic acid into 250ml conical flask.
2- Titrate the $K_2Cr_2O_7$ solution with 0.1N of oxalic acid solution using diphenylamine indicator until the color of the solution becomes purple.
3- Repeat this experiment three times, until three concordant reading are obtained and calculate the amount of potassium dichromate in gram per liter.

3- Calculations

Suppose the required volume of $K_2Cr_2O_7$ consumed for titration process is 10ml; Milliequivalents of oxalic acid = Milliequivalents of $K_2Cr_2O_7$.

$$NV = N1V1$$

$$V_1 (K_2Cr_2O_7) = (0.1x10)/10 = 0.100$$

The amount of $K_2Cr_2O_7 = 0.100 \times 49 = 4.90$ g/l

IV.2 Gravimetric analysis

IV.2.1 Introduction

The gravimetric analysis is the technique depended on the measurement of the mass of a substance of known components, and also it is a process of separating the compounds or elements in pure form. There are three types of the gravimetric analysis methods; they are: volatilization method, electrolysis method and precipitation method. In this book, we will discuss the precipitation method under gravimetric analysis technique.

IV.2.2 Precipitation method

The addition of a salt solution to the other salt solution produces two products, one of them is formed as ppt (insoluble compound) and the other is formed as a solution (soluble compound). To illustrate this point, the addition of potassium chromate solution to a lead nitrate solution gives lead chromate (yellow ppt; $PbCrO_4$; insoluble in water) and a soluble compound of potassium nitrate as shown in the following chemical equation:

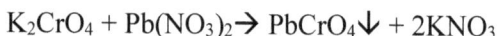

$$K_2CrO_4 + Pb(NO_3)_2 \rightarrow PbCrO_4\downarrow + 2KNO_3$$

Four steps are needed to favor the conditions for precipitation; they are:
1- Precipitation from dilute solutions,
2- Add dilute precipitating reagent slowly with effective stirring,
3- Precipitation at low pH, which is possible to improve the quantitative precipitation. Many ppts are soluble in acidic media which makes the rate of the precipitation slow, and they are more soluble, because the acid radical (anion) combines with the protons in solution,
4- Precipitation from hot solution. This precipitation may be occurred in hot solution.

IV.2.3 Completeness of the precipitation

The completeness of the precipitation is confirmed by waiting until the ppt has settled and then adding few drops of the precipitating agent to the clear solution above it as in the formation of silver chloride AgCl which formed from the reaction of sodium chloride and silver nitrate. The silver chloride ppt is checked by adding few drops of silver nitrate to make sure that the final ppt did not need more reagents. If no turbidity is present, then the reaction is completed.

I- Digestion of the precipitation

Digestion process is usually done at elevated temperature to increase the speed of the process. Sometimes it is done at ambient temperature.

II- Filtration, washing and drying processes

These processes are done after digestion of the ppt on water bath, by using suitable method such as suction by using Buchner tunnel or sintered glass crucible (proper number)]. The formed ppt is isolated from the mother liquor (filtrate), then the ppt washes several times with small volumes of the used solvent until the filtrate becomes clear. The obtained ppt then is dried and weighed.

IV.2.4 Laboratory experiments

I- Estimation of water hydration in the crystals

This experiment aims at the loss of water of hydration above 100°C according to the following chemical reaction.

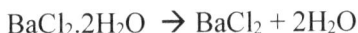

$$BaCl_2.2H_2O \rightarrow BaCl_2 + 2H_2O$$

1- Required chemicals and tools

Crystallized barium chloride, porcelain ignition crucibles, stands, clamps and flames.

2- Procedures

1- Heat the crucible and lid to dull redness for several minutes, and allow it to cool in a desiccators for several minutes, then weigh the crystallized barium chloride after half an hour.
2- Place the covered crucible on silica crucible. Introduce about 2.5g of the salt in the triangle, about 15cm above a small flame (not more than 6-8cm high).
3- Increase the flame gradually until the bottom of the crucible is heated to dull redness. Fix the crucible at this temperature for about 15 minutes, and then allow it to cool in a desiccator for 45 minutes, and weigh the product. Repeat this process until a constant weight obtained.

3- Calculations

From the loss in the mass, we can calculate the percentage of water in barium chloride dehydrate as:
- Weight of crystallized salt (about 1.50g)
- Lost weight = 1.50- lost weight of $BaCl_2$ = w_1 (weight of water of hydration)
- The percentage = $w_1/1.50$ x100

II- Estimation of lead as lead chromate

The main aim of this experiment is to estimate the percentage of lead in lead chromate compound, and due to the difficulty of solubility of chromate, it has a limitation in its applicability, but it is very accurate estimation.

1- Required chemicals and tools

$Pb(NO_3)_2$, K_2CrO_4, CH_3CO_2H, CH_3CO_2Na and distilled water. The tools can be figured out.

2- Procedures

1- Weigh about 0.3g of $Pb(NO_3)_2$, dissolve it in 100ml of distilled water and add few drops of dil. CH_3CO_2H to the solution until it becomes acidic media.
2- Heat the acidic solution to boiling and add from the burette 10ml of 4% of K_2CrO_4 solution. Boil the mixture for about 10 minutes until the yellow ppt settled.
3- Filter the ppt through a sintered glass crucible (No.4). Wash several times with hot dil. CH_3CO_2Na solution then with a hot water. Dry the obtained ppt at 120°C to constant weight.

3- Calculations

To calculate the percentage of lead in lead chromate, we follow the following steps:
- Weight of empty crucible = w_1
- Weight of empty crucible and the ppt = w_2
- Weight of the ppt ($PbCrO_4$) = $w_2 - w_1 = w_3$
- Gravimetric factor of lead in the lead chromate = 207/323 = 0.64087
- Percentage of lead in the sample = w_3 x 0.64087 = (w_4/0.30) x 100

III- Estimation of nickel as nickel dimethylglyoxime

To estimate the purity of nickel in a sample by forming a ppt with dimethylglyoxime in the presence of ammonium hydroxide, The red ppt is washed with a hot distilled water for several times until the color of the filtrate becomes clear and drying it at 120°C and weighed as bis(dimethylglyoximato)nickel(II), as shown in the following chemical equation:

$$NiCl_2 + 2HDMG \rightarrow [Ni(DMG)_2]\downarrow + 2HCl$$

The formed ppt is soluble in alcoholic solutions and mineral acids, but insoluble in the ammonium salt solutions and ammonia solutions.

1- Procedures

1- Weigh about 0.3g of $NiCl_2.6H_2O$ Hexahydrate and transfer it to 500ml beaker.
2- Dissolve the $NiCl_2.6H_2O$ in distilled water, and add 5ml of dilute HCl and complete the volume to 200ml.
3- Heat the solution to 70-80°C, and then add slight excess 20-25ml of dimethyl- glyoxime and immediately add dilute ammonium hydroxide with stirring until the ppt takes place.
4- Allow the ppt to stand on bath water for 15-20 minute, and then test the solution by adding few drops of dimethylglyoxime reagent to complete precipitation when the red precipitation is settled out. Digest the formed precipitation for one hour. Filter it through sintered glass crucible (No.4) and wash it several times with hot water until it is free from chloride ions.
5- Dry the sintered glass crucible in electric oven at 120°C and then allow the crucible to cool in a desiccator and weigh the final product.

2- Calculations

To calculate the percentage of nickel in Bis(dimethylglyoximato) nickel(II), follow the following steps:
- Weight of empty sintered glass crucible = w_1
- Weight of empty crucible and the ppt = w_2
- Weight of the ppt $[Ni(DMG)_2 = w_2 - w_1 = w_3$
- Gravimetric factor of Nickel in the $Ni(DMG)_2$ = 58.7/288.7 = 0.203325
- Percentage of lead in the sample = w_3 x 0.203325 = (w_4/0.30) x 100

IV- Estimation of chloride as silver chloride

To estimate the percentage of chloride content in AgCl, is difficult, because the gravimetric estimation of chloride content is depended on the precipitation of the slightly soluble AgCl. The precipitation process is achieved in the existence of dilute nitric acid to prevent the precipitation of other slightly soluble silver salts such as Ag_3PO_4 or Ag_2CO_3, and the following chemical equation illustrates the precipitation process:

$$NaCl + AgNO_3 \rightarrow AgCl\downarrow + NaNO_3$$

1- Procedures

1- Weigh 0.20g of KCl into 250ml beaker and dilute to 100ml with distilled water and acidify with 2-4 drops of conc. HNO_3.
2- Heat the mixture to 70-80°C and slowly add 2% $AgNO_3$ solution with constant stirring and test the mixture by few drops of $AgNO_3$ until the ppt is completed.
3- Allow the ppt to stand in dark place for two hours, filter it through sintered glass crucible (No: 4), then wash it several times with distilled water and dry it in oven at 120°C. The obtained ppt is cooled and weighed.

2- Calculations

To calculate the percentage of chloride content in $AgNO_3$, follow the following steps.
- Weight of empty sintered glass crucible = w_1
- Weight of empty crucible and the ppt = w_2
- Weight of the ppt [AgCl) = $w_2 - w_1 = w_3$
- Gravimetric factor of Cl content in AgCl = 35.45/143.45= 0.247124
- Percentage of lead in the sample = w_3 x 0.247124 = $(w_4/0.20)$ x 100

V- Estimation of iron and aluminum

Iron can be pptd as ferric hydroxide; Fe (OH) $_3$, while the aluminum remains in the solution as sodium aluminate. Ferric hydroxide is filtered, washed and ignited to ferric oxide; Fe_2O_3. Aluminum can be pptd from the filtrate as white gelatinous precipitation of aluminum hydroxide; $Al(OH)_3$. Firstly, the filtrate is acidified by conc. HCl acid and few grams of ammonium chloride; $AlCl_3$ is added followed by addition of NH_4OH until the mixture becomes faintly alkaline to methyl red indicator. The obtained aluminum hydroxide precipitation is filtered, washed and ignited to alumina compound (Al_2O_3).

1- Procedures

A- Iron

1- Transfer 10ml of the sample solution in 250ml beaker and dilute the solution to 80ml by distilled water.
2- Heat the solution up to 60°C, and then add excess of 0.2N of NaOH slowly with constant stirring. The excess of NaOH will dissolve the aluminum ppt as $NaAlO_2$.
3- Dilute the mixture to 250ml with boiling water, the boil the mixture for several minutes and filter through filter paper (ash less paper). Wash the beaker, ppt and filter with hot water until the filtrates are the most faintly alkaline.
4- Reserve the filtrate and place the ppt paper in ignited crucible, dry and ignite with flame burner to fix weight.

B- Aluminum

1- Add few drops of methyl red indicator to the aluminum ppt (Al_2O_3), then acidify the filtrate with dilute HCl and add 10g of pure NH_4Cl, and heat to boiling, then add dilute NH_4OH solution drop wise until the color of the solution becomes yellow.
2- Boil the solution for 2-3 minutes and digest it on water bath for about one hour, then filter it while hot through ash less filter paper. Wash the formed ppt with hot 2% NH_4NO_3 solution made it neutral with NH_4OH solution to methyl red.
3- Place the ppt paper in crucible and should be ignited to fixed weight.

2- Calculations

Iron: To calculate the percentage of iron content in Fe_2O_3, follow the following steps:
- Weight of empty ignited crucible = w_1
- Weight of empty crucible and the ppt = w_2
- Weight of the ppt Fe_2O_3 = $w_2 - w_1 = w_3$
- Gravimetric factor of Fe content in Fe_2O_3 = 0.69940
- Percentage of Fe in the sample = w_3 x 0.69940 = (w_4/w_{sample}) x 100

Aluminum: To calculate the percentage of aluminum content in Al_2O_3, follow the same as done for iron steps (above).

Questions

1. What are the main principles for the titration of sulphuric acid, sodium carbonate and sodium hydroxide?
2. Calculate the equivalent weights for ; Disodium phosphate, lead chromate, Iron(III) oxide and, Oxalic acid, Ammonium iron(II) sulphate hexahydrate, Copper(I) iodide and Beryllium chloride (Atomic Wts; Sodium =23, phosphor =31, oxygen = 16, lead =207, Chromium =52, Iron= 56, Carbon = 6, Sulphur = 32, Copper = 63.50, Beryllium = 9).
3. Choose the right answer:
   *- The molarity of 0.23N calcium fluoride is (0.23 , 0.46 or 0.115)
   *- The normality of 0.00055M iron(III) sulphate is (0.013, 0.00033 or 0.0033).
   *- The end point color for the titration of sulphuric acid and sodium hydroxide is (red, violet, blue, colorless, none of them).
   *- The color of sodium carbonate salt is (white, colorless, violet, pale green).
   *- The color of the used E.B.T indicator in complexometric titration is (Black, Red, Pink) due to the addition of (Ammonium chloride, Potassium chloride, Sodium chloride) to the original indicator in ratio of (1:1, 1:100 or 0.1:10).
4. A 8ml of potassium dichromate solution was required to titrate 10ml of 0.12N of ammonium iron(II) sulphate hexahydrate solution.
   i) Write the balanced chemical equation for this titration.
   ii) Calculate the strength of the potassium dichromate.
5. Define the following terms: Primary standard solution, Indicator, Titration, Acidity, Normality Salt and Gravimetric factor.
6. By giving examples, state the relationship between molarity and normality.
7. Fill out the following blanks:
   *- The chemical formula of ethylenediaminetetraaceticacid is ....
   *- The chemical structure of EBT is ....
   *- The color of phenolphthalein in acidic media is ....
   *- There are four types of titration, which are ....
   *- The chemical formula of antiacid tablet is ....

*- The molecular weight of aluminum sulphate is ......, whereas its equivalent weight is ......

*- The end point is ....

*- The end point of the hardness water titration is ....

8. Define the main principle of precipitation titration.

9. Write down a brief note on the following; Digestion, filtration, Completeness of precipitate, solvents and centered glass crucible.

10. If 0.400g sample of aluminum containing compound yielded 0.38880g of iron(III) oxide. Perform the following:

   *- Write down the complete chemical reactions.

   *- Find out the percentage of iron and iron(III) oxide in the compound.

11. How do you estimate the cations (Ni^{2+}, Ag^+, Al^{3+} and Pb^{2+}) in their compounds?

Chapter V

Synthesis and Analysis of Some Coordination Compounds

Experiment # 1

Synthesis of bis(acetylacetonato)copper(II)

Introduction

Acetylacetone (2,4-pentanedione; acac) is a typical diketone that can ionize in aqueous solution as a weak acid:

$$CH_3COCH_2COCH_3 \rightleftharpoons H^+ + [CH_3COCHCOCH_3]^-$$

The resulting acetylacetonate anion can serve as a ligand to metal ions, forming complexes in which the ligand is bonded to the metal through both oxygen atoms to form a six-membered ring. The six-membered rings are planar and weakly aromatic because they contain 6 π electrons. In complexes of stoichiometry $Cu(acac)_2$, the $Cu-O_4$ group is square planar. In general, the $M(acac)_2$ complexes are neutral and may be isolated as crystalline solids having an interesting variety of colors.

Chemical reactions

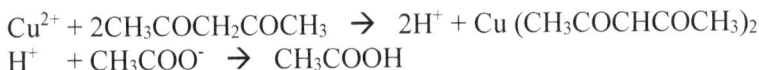

$$Cu^{2+} + 2CH_3COCH_2COCH_3 \rightarrow 2H^+ + Cu(CH_3COCHCOCH_3)_2$$
$$H^+ + CH_3COO^- \rightarrow CH_3COOH$$

Materials required

$CuCl_2.2H_2O$, Acetylacetone, Methanol and Sodium acetate

Procedures

1- Dissolve 0.4g of $CuCl_2.2H_2O$ in 5ml of distilled water.
2- Over a period of 20 min, add drop wise a solution of 0.5ml of acetylacetone in 1ml methanol, with constant stirring.
3- Add 0.68g of sodium acetate dissolved in 1.5ml distilled H_2O over a period of 5min.

4- Heat the mixture to about 80°C on a hot plate for 15min, maintaining rapid stirring.

5- Cool to room temperature, then in an ice bath. Filter the product, wash with 10ml of cold distilled water, and dry it for 15min, and determine % yield.

Recrystallization

1- Place about 0.2g of crude product in a 100 ml Erlenmeyer flask and add 25ml methanol.

2- Place a small glass funnel in the mouth of the flask to serve as a reflux condenser, and boil on a steam bath for 5min.

3- Carefully decant the solution into a 100ml flask, leaving any solid residue behind.

4- Add 5ml of methanol, and reheat to dissolve, and then cool to room temperature.

5- Filter the product, wash with a little ice-cold methanol, suction dry, and then air dry.

Calculations

M.W. of $CuCl_2.2H_2O$ = 170.5g
M.W. of [Cu $(acac)_2$] = 261.5g
170.5g of $CuCl_2.2H_2O$ (reactant) gives 261.5g of [Cu(acac)$_2$] (product)
0.4g of $CuCl_2.2H_2O$ (reactant) gives the weight of [Cu(acac)$_2$] (product)
0.4g /170.5g X 261.5g = 0.61g
% purity of the obtained complex = weight of the product/0.61 x 100

Experiment # 2

Synthesis of hexaamminecobalt(III) chloride complex

Introduction

The cobalt (II) ion is oxidized to cobalt (III) ion by adding H_2O_2. Only few salts of Co(III) such as CoF_3 are known. However, complexation stabilizes the higher oxidation state, and a number of very stable octahedral coordinated complexes of cobalt (III) are known.

Chemical equations

$$Co^{2+} + NH_4^+ + \tfrac{1}{2} H_2O_2 \rightarrow [Co(NH_3)_6]^{3+}$$

$[Co\ (NH_3)_6]^{3+} + 3Cl^- \rightarrow [Co(NH_3)_6]Cl_3$

$2CoCl_2.6H_2O + 2NH_4Cl + NH_3 + H_2O \rightarrow 2[Co(NH_3)_6]Cl_3$

Materials Required

Cobalt (II) chloride hexahydrate, Conc. NH_3, Conc. HCl, 30% H_2O_2, NH_4Cl, 2N NaOH, 2N HNO_3, Ethanol and Potassium iodide.

Procedures

1- Dissolve 12g of NH_4Cl and 18g of $CoCl_2.6H_2O$ in 25ml of hot distilled water.
2- Carefully, add the mixture 1g of charchoal, then cool the mixture in ice bath until 0^0C.
3- Add 40ml of Conc. NH_3 and keep the temperature up to $10°C$.
4- By stirring add to the mixture about 35ml of 30% H_2O_2 at $20°C$. Then heat the mixture to $60°C$.
5- Cool, the mixture in ice bath, then collect the obtained crystals by Buchner funnel and dissolve the crystals in a hot mixture of 160ml of distilled water and 5ml of conc. HCl.
6- Remove the charcoal by filtration and add about 10ml of the concentrated HCl to the filtrate. After cooling, collect the product by a Buchner funnel and wash it by acetone.

Calculations

M.W. of $CoCl_2.6H_2O = 237g$

M.W. of $[Co\ (NH_3)_6]\ Cl_3 = 267.5g$

237g of $CoCl_2.6H_2O$ (reactant) gives 267.5g of $[Co\ (NH_3)_6]\ Cl_3$ (product)

10 g of $CoCl_2.6H_2O$ (reactant) gives Xg of $[Co\ (NH_3)_6]\ Cl_3$ (product)

10g /237g x 267.5g = 11.28g.

% purity of the obtained complex = weight of product / 11.28 x 100

= Xg / 11.28g x 100

Experiment # 2a

Analysis of cobalt ion in $[Co(NH_3)_6]Cl_3$ complex

Chemical equations

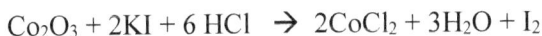

$Co_2O_3 + 2KI + 6\ HCl \rightarrow 2CoCl_2 + 3H_2O + I_2$

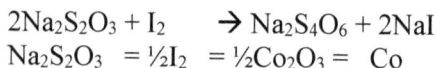

$2Na_2S_2O_3 + I_2 \rightarrow Na_2S_4O_6 + 2NaI$

$Na_2S_2O_3 = \frac{1}{2}I_2 = \frac{1}{2}Co_2O_3 = Co$

Procedures

1- Accurately weigh 0.1g of cobalt complex into a conical flask.
2- Add 20 ml dil NaOH solution and boil the solution until ammonia escapes out.
3- Cool the resulting alkaline suspension of cobaltic oxide and add about 1g of KI.
4- Acidify the liquid with dil HCl (Check with litmus paper – blue litmus turns to red).
5- Add another 40ml of dil HCl to get a clear solution.
6- Fill the burette with 0.1N $Na_2S_2O_3$ solution.
7- Add few ml of freshly prepared starch solution (indicator) to the solution in the conical flask.
8- Titrate until the blue color disappears and pink color (due to formation of cobaltous ion) remains in the flask.
9- Note the volume of $Na_2S_2O_3$ consumed (burette reading) as 'x' ml.

Calculations

1ml of 0.1N $Na_2S_2O_3$ = 0.005894g of Cobalt
xml of 0.1N $Na_2S_2O_3$ = X x 0.005894g of Cobalt (Actual yield)
% Yield of Cobalt in the prepared complex = Actual yield / Theo. yield x 100
Theoretical % of Cobalt in $[Co(NH_3)_5Cl]Cl_2 = 22.0$

Experiment # 2b

Analysis of chloride ion in $[Co(NH_3)_6]Cl_3$ complex

1- Weigh accurately 0.3g of the cobalt complex prepared into a conical flask.
2- Add NaOH (20 ml; 2N) and boil the solution on a Bunsen burner to remove ammonia
3- Cool and filter the solution; Collect the filtrate.
4- Wash the cobaltic oxide ppt with 50ml hot water and combine the washings with the filtrate.
5- Acidify the solution with 2N HNO_3 (Check with litmus paper); Cool the solution.

6- Add 0.1N $AgNO_3$ solution dropwise until the precipitation is complete.

7- Heat the suspension until the ppt coagulates.

8- Collect (filter) the ppt on a G4 sintered glass crucible.

9- Wash the ppt with 0.01N HNO_3 until the washings give no ppt for Ag^+ with HCl.

10- Dry the ppt (AgCl) in an oven (150°C) and allow it to cool in a desiccator and weigh.

11- Repeat the drying, cooling and then weighing until a constant weight is obtained.

Calculations

Gravimetric factor = F.W. of Cl^- / F.W. of AgCl = 35.35 / 143.5 = 0.2474

Experimental %yield = Wt. of the ppt x GF / Wt. of the sample x 100

Experiment # 3

Synthesis of chloropentaamminecobalt(III) chloride complex

Introduction

The cobalt (II) ion is more stable than the cobalt (III) ion for simple salts of cobalt. Only few salts of Co(III) such as CoF_3 are known. However, complexation stabilizes the higher oxidation state, and a number of very stable octahedrally coordinated complexes of cobalt (III) are known.

Chemical equations

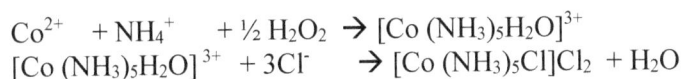

Co^{2+} + NH_4^+ + ½ H_2O_2 → $[Co (NH_3)_5H_2O]^{3+}$

$[Co (NH_3)_5H_2O]^{3+}$ + $3Cl^-$ → $[Co (NH_3)_5Cl]Cl_2$ + H_2O

Materials required

Cobalt(II) chloride hexahydrate, NH_4Cl, 30% H_2O_2, Ethanol, conc. HCl, 2N HNO_3, conc. NH_3, 2N NaOH and Potassium iodide.

Procedures

1- In a fume hood, add 5gm of ammonium chloride to 30 ml concentrated aqueous ammonia in a 250ml Erlenmeyer (Conical) flask. (The combination of NH_4Cl and NH_3 (aq) guarantees a large excess of the NH_3 ligand.)

2- Stir the ammonium chloride solution vigorously using a magnetic stirring plate while adding 10g finely divided $CoCl_2.6H_2O$ in small portions.

3- Next, add 8ml 30% hydrogen peroxide to the brown Cobalt slurry, using a burette that has been set up in the hood and filled by the laboratory instructor. An addition rate of about 2 drops per second is usually sufficient, but care should be taken to avoid excessive effervescence in this exothermic reaction. (If the reaction shows signs of excessive effervescence, turning off the magnetic stirrer momentarily will usually prevent overflow of the solution.)

4- When the effervescence has ceased, add 30 ml conc. HCl with continuous stirring, pouring 1-2ml at a time. At this point, the reaction may be removed from the hood.

5- Use a heater to heat the solution and hold the temperature between 55-65°C for 15min with occasional stirring. This incubation period is necessary to allow complete displacement of all aqua ligand.

6- Add 25 ml distilled water, and allow the solution to cool to room temperature.

7- Collect the purple product by filtration through a Buchner funnel; wash it three times with 7.5ml cold distilled water and twice with 7.5ml ice-cold ethanol. The solutions must be cold to prevent undue loss of product by re-dissolving.

8- Transfer the product to a crystallizing dish, loosely cover with aluminium foil, and allow drying until the following laboratory period. record the yield.

Calculations

M.W. of $CoCl_2.6H_2O = 237g$
M.W. of $[Co (NH_3)_5Cl]Cl_2 = 250.5g$
237g of $CoCl_2.6H_2O$ (reactant) gives 250.5g of $[Co(NH_3)_5Cl]Cl_2$ (product)
10g of $CuSO_4.5H_2O$ (reactant) gives Xg of $[Cu(NH_3)_4]SO_4.H_2O$ (product)

10g / 237g x 250.5g = 10.57g
Experimental yield of the complex = Xg (weight of the synthesized complex).
% purity of the obtained complex = weight of product / 10.57 x 100
= Xg / 10.57g x 100

Experiment # 4

Synthesis of tetraamminecopper(II) sulphate hydrate complex

Introduction

In aqueous solution, typical cations, particularly those produced from atoms of the transition metals, do not exist as free ions but rather consist of the metal ion in combination with some water molecules. Such cations are called complex ions. The water molecules, usually two, four, or six in number, are bound chemically to the metallic cation, but often rather loosely with the electrons in the chemical bonds being furnished by one of the unshared electron pairs from the oxygen atoms in the H_2O molecules. Copper ion in aqueous solution may exist as $[Cu (H_2O)_4]^{2+}$ with the water molecules arranged in a square around the metal ion at the center. If a hydrated cation such as $[Cu(H_2O)_4]^{2+}$ is mixed with other species that can, like water, form coordinate covalent bonds with Cu^{2+}, those species, called ligands, may displace one or more H_2O molecules and form other complex ions containing the new ligands. For instance, NH_3, a reasonably good coordinating species, may replace H_2O from the hydrated copper ion, $[Cu(H_2O)_4]^{2+}$, to form $[Cu(H_2O)_3NH_3]^{2+}$, $[Cu(H_2O)_2(NH_3)_2]^{2+}$, $[Cu(H_2O)(NH_3)_3]^{2+}$, or $[Cu(NH_3)_4]^{2+}$. At moderate concentrations of NH_3, essentially all the H_2O molecules which coordinate to the copper ion will be replaced by NH_3 molecules, forming the copper ammonia complex ion $[Cu(NH_3)_4]^{2+}$.

Chemical equation

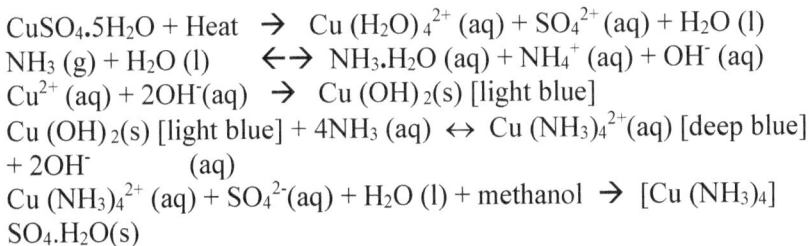

$CuSO_4.5H_2O$ + Heat → Cu $(H_2O)_4{}^{2+}$ (aq) + $SO_4{}^{2+}$ (aq) + H_2O (l)
NH_3 (g) + H_2O (l)   ←→ $NH_3.H_2O$ (aq) + $NH_4{}^+$ (aq) + OH^- (aq)
Cu^{2+} (aq) + $2OH^-$(aq) → Cu $(OH)_2$(s) [light blue]
Cu $(OH)_2$(s) [light blue] + $4NH_3$ (aq) ↔ Cu $(NH_3)_4{}^{2+}$(aq) [deep blue]
+ $2OH^-$ (aq)
Cu $(NH_3)_4{}^{2+}$ (aq) + $SO_4{}^{2-}$(aq) + H_2O (l) + methanol → $[Cu (NH_3)_4]$ $SO_4.H_2O$(s)

Materials required

Copper(II) sulphate pentahydrate ($CuSO_4.5H_2O$), Conc. NH_3, Methanol, Potassium Iodide (10% Solution), 0.1N Sodium thiosulphate($Na_2S_2O_3$) solution, Freshly prepared Starch solution, Conc. HCl, Barium chloride($BaCl_2$; 5% Solution), Silver nitrate($AgNO_3$ Solution) and Acetic acid.

Procedures

1- Weigh accurately 7g of copper(II) sulphate pentahydrate into a conical flask.
2- Add 15ml distilled water and stir well with a glass rod.
3- Boil the contents gently over a burner until entire solid has dissolved.
4- Cool the solution to room temperature. Now Cu^{2+} ions are in the aqueous phase.
5- Under the hood, add 5–7ml of concentrated ammonia one ml at a time.
6- Gently swirl the flask until the blue ppt dissolves (If there is anymore ppt, one more ml of ammonia and swirl again).
7- Add about 15 ml ethanol and gently swirl the solution until the deep blue of tetraammine copper(II) sulphate complex formed.
8- Allow the mixture to stand in ice bath for 3-4 minutes.
9- Filter the ppt and wash it with 1-2ml aliquots of ethanol 2-3 times.
10- Allow the ppt. to air dry for about 25 minutes and record the yield of the complex.

Calculations

M.W. of $CoCl_2.6H_2O = 237g$
M.W. of $CuSO_4.5H_2O = 249.5g$
M.W. of $[Cu (NH_3)_4]SO_4.H_2O = 245.5g$
249.5g of $CuSO_4.5H_2O$ (reactant) gives 245.5g of $[Cu(NH_3)_4]SO_4.H_2O$ (product)
7g of $CuSO_4.5H_2O$ (reactant) gives Xg of $[Cu(NH_3)_4]SO_4.H_2O$ (product)
7g /249.5g x 245.5g = 6.89g
Experimental yield of the complex = Xg (weight of the synthesized complex).
% purity of the obtained complex = Experimental Yield / Theoretical Yield x 100
= (Xg / 6.89g) x 100

Experiment # 4a

Analysis of copper ion in [Cu(NH₃)₄]SO₄ complex

Chemical equations

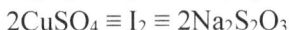

$2CuSO_4 + 4KI \rightarrow 2CuI + I_2 + 2K_2SO_4$

$2S_2O_3^{2-} + I_2 \rightarrow S_4O_6^{2-} + 2I^-$

$2CuSO_4 \equiv I_2 \equiv 2Na_2S_2O_3$

Procedures

1- Weigh accurately 0.5g of complex and dissolve in 50ml dilute HCl in a conical flask
2- Add conc. NH_3 drop wise until a faint green ppt of copper hydroxide is formed
3- Add 2-3 drops of dilute acetic acid until the ppt just dissolves
4- Add about 10ml of 10% KI solution until brown color appeared.
5- Shake and allow the brown solution to stand for 5 minutes
6- Fill burette with 0.1N sodium thiosulphate solution.
7- Titrate until a faint yellow color appears, and add about 1ml of freshly prepared starch solution; color changes to blue.
9- Continue the titration by adding sodium thiosulphate solution drop by drop until a blue color changes to white (end point).
10- Record the burette reading i.e. volume of sodium thiosulphate as 'X' ml.

Calculations

1ml of 0.1N sodium thiosulphate = 0.00635g of Cu^{2+}
Xml of sodium thiosulphate = X x 0.00635 =_____ g of Cu^{2+}
% yield of Cu^{2+} =
Theoretical % Cu in the complex =
Atomic wt. of the Cu (II) x y / M.W. of the $[Cu(NH_3)_4]SO_4.H_2O$ x100
Where y = No. of copper ions
63.5 x 1 / 245.5g x100 = 25.86
Experimental % of the Cu (II) = V x N x Eq. Wt. of the Cu(II) / (Wt. of the sample x 1000) x 100
Where V = Volume of the titrant (hypo) or burette reading
N = Normality of the titrant
E.g. 9ml x 0.1 x 63.5 / (0.2g x 1000) x100 = 23.3%
% Purity = (Expt. / Theo.) x100 = (23.3 / 25.86) x 100 = 90.1

Experiment # 4b

Analysis of sulphate ion in $[Cu(NH_3)_4]SO_4$ complex

Chemical equation

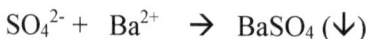

$SO_4^{2-} + Ba^{2+} \rightarrow BaSO_4 (\downarrow)$

Procedure

1- Dissolve 0.4g of the complex it in 150ml distilled water and add 2ml of conc. HCl. Then add 10ml 5% $BaCl_2$ solution while stirring

2- Cover the beaker with a watch glass and leave it on water-bath for one hour.

3- Remove the beaker from the water-bath and test the supernatant clear liquid for complete precipitation of $BaSO_4$ by adding few drops of $BaCl_2$ solution.

4- If no fresh precipitation occurs, collect the ppt on a previously weighed and dried sintered glass crucible (No.4), wash the ppt with warm water until the filtrate gives no ppt with $AgNO_3$ solution.

8- Dry the sintered glass crucible containing the ppt at 110°C for one hour.

9- Cool it in the desiccators and weigh.

10- Repeat the process until a constant weight is obtained

11- Record the weight of the ppt as 'X'g

Calculations

1g of $BaSO_4$ = 0.4114g of SO_4^{2-}

x g of $BaSO_4$ = _____ g of SO_4^{2-}

% yield of SO_4^{2-} =

Gravimetric Factor = F.W. of SO_4^{2-}/ F.W. of $BaSO_4$ = 96/233.3 = 0.4115

Experimental % = (Wt. of the ppt x GF) / Wt. of the sample x 100

Experiment # 5

Synthesis of hexaamminenickel(II) chloride complex

Introduction

Nickel(II) ion forms salts with virtually every anion and has an extensive aqueous chemistry based on the green ion $[Ni(H_2O)_6]^{2+}$, which is always present in the absence of strong ligands. Octahedral complexes of Ni (II) produced often from aqueous solution by replacement of coordinated water, especially with neutral N-donor ligands such as NH_3.

Materials required

Nickel(II) chloride hexahydrate, Conc. Ammonia, 95% Ethanol, Conc. HCl, Alcoholic dimethyglyoxime, ammonium hydroxide solution and 0.1N NaOH.

Chemical equation

$NiCl_2.6H_2O + 6NH_3 \rightarrow [Ni(NH_3)_6]Cl_2 + 6H_2O$

Procedures

1- Weigh accurately 12 g of $NiCl_2.6H_2O$ (green) (M.W. 237.7g) in a 250ml beaker.
2- Dissolve in 20ml warm water + 25ml conc. ammonia under the hood.
3- Cool the deep violet solution in ice. A heavy crystalline of product separates out.
4- Add about 100ml of cold 95% alcohol to the mixture to complete the deposition; the supernatant liquid should be almost colorless.
5- Filter and wash the ppt (violet) with two 25ml portions of alcohol and dry in air.
6- Record the yield of $[Ni(NH_3)_6] Cl_2$ (M.W. 231.7g).
7- Theoretical yield = 11.70g $\equiv$ 6g $NiCl_2.6H_2O$.

Calculations

% yield of hexamminenickel (II) Chloride = _____ g
M.W. of $NiCl_2.6H_2O$ = 237.7g

M.W. of $[Ni(NH_3)_6]Cl_2$ = 231.7g
237.7g of $NiCl_2.6H_2O$ (reactant) gives 231.7g of $[Ni(NH_3)_6]Cl_2$ (product)
12g of $NiCl_2.6H_2O$ (reactant) gives ? Xg of $[Ni(NH_3)_6]Cl_2$ (product)
12g / 237.7g X 231.7g = 11.70g
Experimental yield of the complex = Xg (weight of the synthesized complex).
% purity of the obtained complex = Experimental Yield / Theoretical Yield x 100
= (Xg/11.70g) x 100

Experiment # 5a

Analysis of nickel ion in $[Ni(NH_3)_6]Cl_2$ complex

Chemical equation

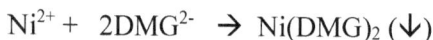

Ni^{2+} + $2DMG^{2-}$ → $Ni(DMG)_2$ ($\downarrow$)

Procedures

1- Weigh accurately 0.2g of the complex in a 500ml beaker. Dissolve the complex in 2.5ml conc. HCl, add 100ml distilled water and then heat up to 75°C.
2- Add approximately 20ml alcoholic dimethylglyoxime $C_4H_7N_2O_2$, and then add approximately 20 drops of NH_4OH while stirring until precipitation occurs.
3- Add approximately 10 drops more NH_4OH solution to the ppt, and heat the mixture on water-bath for 25min, a red ppt is formed
4- Test the complete precipitation by adding few more drops of dimethylglyoxime.
5- Allow the ppt to settle at room temperature for one hour.
6- Filter the ppt through a previously weighed and dried sintered glass crucible (No.4).
7- Wash the ppt with cold water until the washings are free from chloride ion
8- Dry the ppt at 100-120°C for 60min, and cool the ppt of bis(dimethyl glyoximato) nickel(II), $[Ni(C_4H_7O_2N_2)_2]$ in desiccator and weigh.
9- Repeat the drying and weighing until a constant weight is obtained.

Calculations

FW of the complex = 288.7g/mol

1g of bis(dimethylglyoximato) nickel(II) = 0.2033g of Ni^{2+}

% yield of nickel (II) = _____ g

Gravimetric Factor = F.W. of Ni^{2+} / F.W. of [Ni(DMG)$_2$] = 58.7 / 288.7

= 0.2038

Experimental % = (Wt. of ppt x GF / Wt. of sample) x 100

Experiment # 5b

Analysis of ammonia in [Ni(NH$_3$)$_6$]Cl$_2$ complex

Chemical equation

$NH_4^+ + OH^- \rightarrow NH_3 (\uparrow) + H_2O$

Procedures

1- Weigh out accurately 0.2g of the complex into a 500 ml conical flask, and add 100 ml of standard 0.1N hydrochloric acid solution.

2- Place a small funnel in the neck of the flask in order to prevent mechanical loss, and boil the mixture until a piece of filter paper moistened with mercury(I) nitrate solution and held in the escaping steam is no longer turned black.

4- Cool the solution and add a few drops of methyl red indicator.

5- Fill burette with 0.1N sodium hydroxide solution.

6- Titrate until the end point and record the burette reading as 'y' *ml*

Calculations

1ml of 0.1N HCl = 1 ml of 0.1N NaOH = 0.0017g of NH$_3$

yml of 0.1N NaOH = y x 0.0017 = _____ g of NH$_3$

% yield of NH$_3$ = _____

Experiment # 5c

Analysis of chloride ion in [Ni(NH$_3$)$_6$]Cl$_2$ complex

Chemical equation

$$Cl^- + AgNO_3 \rightarrow AgCl(\downarrow) + NO_3^-$$

Procedures

1- Weigh accurately 0.3g of the nickel complex into a conical flask, and add 20ml 2N NaOH solution. Then boil the solution on a Bunsen burner to remove ammonia.
2- Cool and filter the solution; Collect the filtrate.
3- Wash the ppt with 50ml hot distilled water and combine the washings with the filtrate.
4- Acidify the solution with 2N HNO$_3$ (Check with litmus paper); Cool the solution.
5- Add 0.1N AgNO$_3$ solution drop wise until the precipitation is complete, and heat it until the ppt coagulates.
6- Collect (filter) the ppt on a G4 sintered glass crucible.
7- Wash the ppt with 0.01N HNO$_3$ until the washings give no ppt for Ag$^+$ with HCl.
8- Dry the ppt (AgCl) in an oven at 150°C and allow it to cool in desiccator, then weigh.
9- Repeat the drying, cooling and weighing until a constant weight is obtained.

Calculations

1g of AgCl = 0.2474g of Cl$^-$
% yield of Cl in the Complex =.....................

Experiment # 6

Synthesis of Potassium trioxalatoferrate(III) trihydrate complex

Introduction

The synthesis of potassium trioxalatoferrate(III), K$_3$[Fe(C$_2$O$_4$)$_3$].3H$_2$O involves two steps. The product of the first step is ferrous oxalate FeC$_2$O$_4$ 2H$_2$O, which is made by heating a solution of Iron(II) ammonium sulphate, (NH$_4$)$_2$ Fe (SO$_4$)$_2$ 6H$_2$O, and oxalic acid, H$_2$C$_2$O$_4$.

In the second step of the synthesis the Iron(II) oxalate from the first step is treated with hydrogen peroxide, H_2O_2, in the presence of oxalic acid and potassium oxalate, $K_2C_2O_4$, to give the complex ion $[Fe(C_2O_4)_3]^{3-}$ (the oxalates are tightly bound to the Fe^{3+}), which precipitated out of solution as the potassium salt, $K_3Fe(C_2O_4)_3.3H_2O$.

Materials required

Iron(II) ammonium sulphate, $(NH_4)_2Fe(SO_4)_2.6H_2O$, 1M oxalic acid, $H_2C_2O_4.2H_2O$, saturated $K_2C_2O_4$ solution (30g $K_2C_2O_4.H_2O$ /100 ml $H_2O=1.8M$), 30% H_2O_2 (oxidizer) 30 g $H_2O_2/100$ ml soln.), 95% Ethanol, 0.1N $KMnO_4$, Zinc dust.

Chemical equations

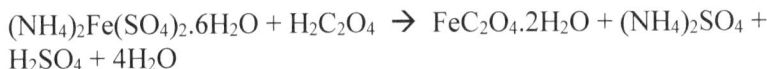

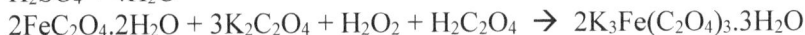

$(NH_4)_2Fe(SO_4)_2.6H_2O + H_2C_2O_4 \rightarrow FeC_2O_4.2H_2O + (NH_4)_2SO_4 + H_2SO_4 + 4H_2O$

$2FeC_2O_4.2H_2O + 3K_2C_2O_4 + H_2O_2 + H_2C_2O_4 \rightarrow 2K_3Fe(C_2O_4)_3.3H_2O$

Procedures

1- Weigh out about 3g of Iron(II) ammonium sulphate on the electronic balance. Put the crystals in the 150ml beaker and add 25 ml of distilled water plus few drops of dil. H_2SO_4. Stir well to dissolve.

2- Now add 25ml of 1.0M oxalic acid to the solution. Put the beaker on a hot plate and heat to boiling, stirring the mixture constantly.

3- Cool the mixture and allow the ppt of ferrous oxalate to settle.

4- Decant the liquid and wash the ppt with 20ml of distilled water.

5- Warm the mixture to 40°C and then allow the ppt to settle again. Decant the wash liquid, removing as much as possible.

6- Add 10ml of saturated potassium oxalate solution to the ppt and warm the mixture to 40°C. Use a ring stand to hold the thermometer.

7- Slowly add 20ml of 30% hydrogen peroxide through burette, stirring continuously while maintaining the temperature at 40°C (if the temperature rises above 40°C during this stage you must allow it to cool before adding the rest of the H_2O_2).

8- When all the hydrogen peroxide has been added, heat the mixture to boiling.

9- Add 5ml of 1.0M oxalic acid all at once, and then add 3ml more dropwise. Keep the mixture near boiling. If the solution is not a clear green color, add up to 2ml more oxalic acid until a clear green is obtained. Continue boiling until the volume is reduced to 25-30ml.

10- Turn off your hot plate and allow the solution to cool, and then place it in an ice bath.

11- After 10min, add 5ml of cold ethanol to aid the crystallization.

12- Allow the mixture undisturbed for additional 15min while crystals are forming.

13- Cool the solution in an ice bath for 10-20 min, then filter, dry and weigh it.

14- Repeat the process until a constant weight is obtained.

15- Record the yield.

Note: The complex is sensitive to light and when exposed to light a yellow powder of Iron (II) oxalate is formed on the surface of the complex. Hence this complex is stored in dark.

Calculations

$2K_3[Fe(C_2O_4)_3] \rightarrow 2FeC_2O_4 + 2CO_2 + 3K_2C_2O_4$

% yield of Potassiumtrioxalato ferrate(III) complex = _____ g

M.W. of $(NH_4)_2Fe(SO_4)_2.6H_2O = 391g$

M.W. of $K_3[Fe(C_2O_4)_3].3H_2O = 491g$

391g of $(NH_4)_2 Fe(SO_4)_2.6H_2O$ (reactant) gives 491g $K_3[Fe(C_2O_4)_3].3H_2O$ of (product)

3g of $(NH_4)_2 Fe(SO_4)_2.6H_2O$ (reactant) gives Xg of $K_3[Fe(C_2O_4)_3].3H_2O$ (product)

3gm /391g x 491g = 3.77g

Experimental yield of the complex = X gm (weight of the synthesized complex).

% purity of the obtained complex = (Experimental Yield / Theoretical Yield) x 100

= (Xg / 3.77g) x 100

Experiment # 6a

Analysis of oxalate ion in $K_3[Fe(C_2O_4)_3].3H_2O$ complex

Chemical equation

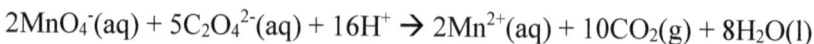

$2MnO_4^-(aq) + 5C_2O_4^{2-}(aq) + 16H^+ \rightarrow 2Mn^{2+}(aq) + 10CO_2(g) + 8H_2O(l)$

Procedures

1- Weigh accurately 0.2g of the complex in a conical flask
2- Add 25 ml 2N H_2SO_4 and 1ml conc.H_3PO_4
3- Heat the solution to boiling (~ 80°C)
4- Fill up the burette with 0.1N $KMnO_4$
5- Titrate until a permanent pink color (end point) is obtained
6- Record the burette reading (i.e. volume of $KMnO_4$ consumed) as 'X'ml

Calculations

1ml of 0.1N $KMnO_4$ solution = 0.0044g of $C_2O_4^{2-}$
'X' ml of 0.1N $KMnO_4$ solution = X x 0.0044 = _____ g of $C_2O_4^{2-}$
% yield of $C_2O_4^{2-}$ = _____

Experiment # 6b

Analysis of iron (III) ion in $K_3[Fe(C_2O_4)_3].3H_2O$ complex

Chemical equation

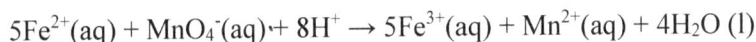

$$5Fe^{2+}(aq) + MnO_4^-(aq) + 8H^+ \rightarrow 5Fe^{3+}(aq) + Mn^{2+}(aq) + 4H_2O\ (l)$$

Procedure: There are two methods to determine the quantity of iron in the complex

First method

1) Add 2g of zinc dust to the oxalate ion analysis and boil for 25 minutes
2) Filter the solution through sintered glass crucible (No.3)
3) Titrate the filtrate solution against 0.1N $KMnO_4$ until the end point becomes pink.
7) Record the burette reading (i.e. volume of $KMnO_4$ consumed) as 'y' *ml*

Calculations

1ml of 0.1N $KMnO_4$ solution = 0.0055g of iron
'y' ml of 0.1N $KMnO_4$ solution = y X 0.0055 = _____ g of Iron
% yield of iron = _____

Second method

1- Weigh about 0.8g of complex into a small porcelain basin.
2- Add 2-3ml of conc. H_2SO_4 and cover the basin with a watch-glass
3- Heat gently for few minutes until the effervescence resulting from the decomposition of the oxalic acid ceases
4- Allow the basin and the contents to cool and then transfer into a conical flask
5- Wash the watch-glass and basin with distilled water and add the washings to the conical flask
6- Heat the turbid liquid until a clear pale yellow solution is obtained
7- Reduce this solution with amalgamated zinc

Preparation of amalgamated zinc

1- Add 2g of pure zinc pellets to a solution of 0.5g $HgCl_2$ in 20ml of 2N HCl
2- Shake the contents for a short time to complete amalgamation
3- Wash the amalgamated zinc under running water
8- A brisk effervescence takes place as soon as amalgamated zinc is added
9- After 20-30min, the solution turns colorless or faint bottle green
10- Titrate the solution against 0.1N $KMnO_4$ as before and record the end point 'y' *ml*

Calculations

1ml of 0.1N $KMnO_4$ solution = 0.0055g of iron
'y' ml of 0.1N $KMnO_4$ solution = y X 0.0055 = _____ g of iron
% yield of iron = _____

Experiment # 7

Synthesis of potassium trioxalatochromate(III) trihydrate complex

Introduction

Oxalic acid, (1,2-ethanedioic acid), in the form of the dianion, functions as a bidentate ligand with many transition metal ions. Chromium is an abundant element in the earth's crust. The metal is used for plating and in chrome steels. The chromates [chromium(VI)] have many industrial uses as pigments, catalysts and fungicides. Chromium(III) is a common and stable oxidation state that displays significant kinetic inertness. The most common geometry is octahedral with other shapes being quite rare.

Materials required

Oxalic acid, Potassium dichromate, Potassium oxalate, Ethanol, $AgNO_3$ solution, Potassium persulphate($K_2S_2O_8$), diphenylamine, iron(II) ammonium sulphate.

Chemical equations

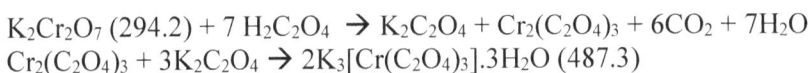

$K_2Cr_2O_7$ (294.2) + 7 $H_2C_2O_4$ → $K_2C_2O_4$ + $Cr_2(C_2O_4)_3$ + $6CO_2$ + $7H_2O$
$Cr_2(C_2O_4)_3$ + $3K_2C_2O_4$ → $2K_3[Cr(C_2O_4)_3].3H_2O$ (487.3)

Procedure

1- Dissolve about 10g of oxalic acid in 20 ml of distilled water
2- Add about 3g of Potassium dichromate in small portions (exothermic reaction!)
3- When the vigorous reaction is subsided, heat the solution to boiling
4- Add about 4g of potassium oxalate ($K_2C_2O_4.H_2O$) to the boiling solution
5- Allow the solution to cool at room temperature, and then add 5ml ethanol. A blue green crystalline formed; $K_3[Cr(C_2O_4)_3].3H_2O$ appears from a blank solution.
6- Allow the ppt to settle for some time and collect the crystalline mass on a Buchner funnel, and wash the ppt first with about 10ml of 1:1 mixture of ethanol and water; and finally with about 5ml of ethanol.
7- Dry at room temperature in the air and record the yield.

Calculations

M.W. of $K_2Cr_2O_7$ = 294g
M.W. of $K_3[Cr(C_2O_4)_3].3H_2O$ = 487g
294g of $K_2Cr_2O_7$ (reactant) gives 487g $K_3[Cr(C_2O_4)_3].3H_2O$ of (product)
3 gm of $K_2Cr_2O_7$ (reactant) gives ? Xg of $K_3[Cr(C_2O_4)_3].3H_2O$ (product)
3g /294g X 487g = 4.97g

Experimental yield of the complex =Xg (weight of the synthesized complex).

% purity of the obtained complex = Experimental Yield / Theoretical Yield x 100

= Xg / 4.97g x 100

Experiment # 7a

Analysis of chromium(III) ion in K₃[Cr(C₂O₄)₃].3H₂O complex

Chemical equation

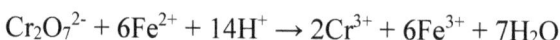

$$Cr_2O_7^{2-} + 6Fe^{2+} + 14H^+ \rightarrow 2Cr^{3+} + 6Fe^{3+} + 7H_2O$$

Procedures

1- Dissolve 0.5g of the complex in 100ml of 2N H_2SO_4
2- Add about 5ml of 0.1N $AgNO_3$ solution and 2g of $K_2S_2O_8$.
3- Boil until solution has the orange red color. (Cr^{3+} ion in the solution is oxidized to $Cr_2O_7^{2-}$ ion. Ag^+ catalyses this oxidation reaction).
4- Cool the solution at room temperature, and then add 50ml of 0.1N ferrous ammonium sulphate solution.
5- Titrate the excess of Fe^{2+} ion against 0.1N potassium dichromate solution using diphenylamine as indicator. Standardized the iron (II) ammonium sulphate (0.1N) with 0.1N potassium dichromate + 1ml 1:1 H_3PO_4: H_2SO_4 (Endpoint: purple).

Calculations

% Experimental yield of Chromium =____g
(0.1 X (50 – burette reading) x 0.001734g Cr^{3+}/ 0.5)x 100
1ml of Fe^{2+} = 0.001734g Cr^{3+}
% Purity= (Experimental value /Theoretical value) x 100

Experiment # 8

Measuring the conductivity of complexes (verification of Werner's theory)

Introduction

This kind of experiments is to verify the Werner's theory by checking the conductivity of some complex compounds. In ionic compounds of the main group elements, it is usually a trivial matter to deduce the number of ions per mole present in infinitely dilute solution. The ionic compounds are viewed as dissociating completely in the dilute solution (although as the concentration of solute rises, the degree of ionization changes drastically), and thus $Ca(NO_3)_2$ would be expected to consist of three ions: one Ca^{2+} and two nitrate (NO_3^-) ions.

In transition metal complexes, the situation is not nearly as simple. A given anion may be a part of the complex (in which case it generally does not dissociate) or it may be presented as a counter-ion (in which case it does). Werner, in 1912, investigated the octahedral complex $[Co(H_2O)_6Cl_3$ which have different potential ligand arrangements in aqueous solution:

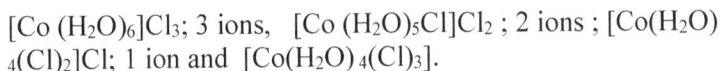

$[Co (H_2O)_6]Cl_3$; 3 ions, $[Co (H_2O)_5Cl]Cl_2$; 2 ions ; $[Co(H_2O)_4(Cl)_2]Cl$; 1 ion and $[Co(H_2O)_4(Cl)_3]$.

The number of ions constituting the complex is best determined by measuring the conductivity of the solution of that compound. This conductivity measurement allows one to tell how many ions (cations and anions) are present in solution when an ionic product is dissolved in water. Those ionic compounds that are soluble in water and conduct electric current in aqueous solution are called electrolytes. The dissolution process consists of complete dissociation of ionic compounds into mobile cations and anions. There are many compounds, which though soluble in water, do not exhibit any conductivity. These are termed non-electrolytes. There is still another group of compounds that exhibit conductance in solutions only when that solution is quite dilute. Such compounds are known as weak electrolytes. Solutions that contain large numbers of mobile ions (cations and anions from the soluble ionic compounds) conduct current well, and solutions that contain only a few ions (acetic acid) or relatively immobile ions show poor conductivity. The conductivity of a solution varies with the number, size, and charge of the ions constituting the solution. The viscosity of a solution also affects the conductivity, by affecting the mobility of the ions. Ions of different species in solution will therefore show different conductivities. If, by means of a chemical reaction, we replace one ionic species by another having a different size and/or charge, we would observe a corresponding change in conductivity of the resulting solution. The conductivity, L, of a solution is presented by the equation; $L = B\ c_i$ $\alpha_i\ Z_I$. Where B is a constant that depends on the size and the geometry of the conductance cell, c_i is the concentration of individual ions in solution, α_i is the equivalent ionic conductance of individual ions, and Z_i is the charge of the ions. In practice, although the conductance of a solution is more useful in dealing with electrolyte solutions, it is the resistance of a solution that is experimentally measured. The conductance is calculated from the resistance. The resistance of a solution is determined by inserting two electrodes into a solution. The resistance, R, is proportional to the distance, d, between the two electrodes and inversely proportional to the cross-sectional area A, of the solution enclosed between the electrodes. i.e. $R = \rho.d\ /\ A$. The term ρ is called the specific resistance or more simply, the resistivity. The ratio d/A is usually referred to as the cell constant, K. Thus the above relation becomes. i.e. $R = K.d$. The conductance, L, of a solution is defined as the reciprocal of the specific resistance.

$$k = 1/\rho = (d\ /\ A)(1/R) = K.L$$

In practice, the cell constant, K, is determined for any cell by measuring the conductivity of a 0.0200M KCl solution at 25°C, for which the specific conductivity, k, is 0.002 768 ohm^{-1}. The total conductivity of a solution arises from several sources, the largest of which is the ions. The self-ionization of a solvent contributes as well, but in practice is small enough to be neglected in all but the most careful measurements. A very useful quantity is the equivalent conductivity; Λ. It is defined as the value of the specific conductivity; k, contributed by one equivalent of ions of either charge. More specifically, it is defined as the conductance of a solution containing one gram-equivalent of an electrolyte placed between electrodes separated by a distance of 1 cm. If c is the concentration of the solution in gram-equivalents/litre and the volume of the solution in cubic centimetres/ equivalent (cm^3/equiv) is equal to 1000/c. The equivalent conductance; Λ, is then given by

$$\wedge = \frac{1000.k}{c}$$

Substituting for k,

$$\wedge = \frac{1000.LK}{c}$$

Another frequently used quantity in conductance measurements is the **molar conductance**; Λ_m which defined as the conductance of a one cubic centimetres volume of solution that contains one mole (or formula weight) of the electrolyte. If M is the concentration of the solution in moles per litre, then the volume in cubic centimetres per mole is 1000/M. The molar conductance is then given by

$$\wedge_m = \frac{1000.k}{M}$$

By comparing the molar conductance measured for a particular compound with that of a known ionic compound, we can estimate the number of ions produced in a solution. A range of values of molar conductance for 2-5 ions at 25°C in water is given below:

No. of Ions	Molar Conductance, cm^{-1}mol^{-1}ohm^{-1}
2	118 – 131
3	235 – 273
4	408 – 435
5	~560

The equivalent conductivity increases with increasing dilution due to the lessened inter-ionic forces between ions (less ion pairing is the classical way of stating this).

Procedures

Conductivity measurements require the use of two instruments: the conductivity cell and the conductivity meter or bridge. The cell constant; K, is first determined by measuring the conductivity; L, of an accurately prepared 0.0200M KCl solution for which the value of specific conductivity; k, is known to be 0.002768 ohm^{-1}, where K = k/L. The molar conductivity of any compound can be determined in the following way.

1- A 1×10^{-3} M solution of the compound of interest is prepared.
2- The conductivity is measured in the same cell for which the cell constant has been previously determined, as described previously.
3- The cell should be rinsed before each measurement.

Experiment # 9

Measuring the potentiometric of complexes

Introduction

This is to check the end point of a titration without using any indicator. There are a lot of quantification methods to find the ions to be found first and second effect area in the coordination compounds. Generally argentometric titrations are preferred due to the chloride ions in the coordination compounds. However, the indicator such as CrO_4^{2-} or florescence should be used in the determination of chloride ions. CrO_4^{2-} indicator changes the color to the tile red, and florescent indicator changes the color to the rose pink at the end of the titration. Due to the colorful solutions of complexes, determination of the end point of a titration is difficult.

Therefore, an instrument should be used to obtain the end point. Potentiometer will be used for the purpose of the titration, due to its applicability. Potentiometric titrations provide data that are more reliable than data from titrations that use chemical indicators and are particularly useful with colored or turbid solutions and for detecting the presence of unsuspected species. *A potentiometric titration involves measurement of the potential of a suitable indicator electrode as a function of titrant volume.* The information provided by a potentiometric titration is not the same as that obtained from a direct potentiometric measurement. Several methods can be used to determine the end point of a potentiometric titration. Refer to Index 3.

The most straightforward method involves a direct plot of potential as a function of reagent volume, as in Figure 1(a) the midpoint in the steeply rising portion of the curve is estimated visually and taken as the end point. Various graphical methods have been proposed to aid in the establishment of the midpoint, but it is doubtful that these procedures significantly improve its determination. A second approach to end-point detection is to calculate the change in potential per unit volume of titrant (that is, $\Delta E/\Delta V$), as in column 3 of Table-1.

A plot of these data as a function of the average volume V produces a curve with a maximum that corresponds to the point of inflection (Figure 1(b)). Alternatively, this ratio can be evaluated during the titration and recorded in lieu of the potential. Inspection of column 3 of Table-1 reveals that the maximum is located between 24.30 and 24.40ml; 24.35ml would be adequate for most purposes. Column 4 of Table-1 and Figure 1(c) show that the second derivative for the data changes sign at the point of inflection. This change is used as the analytical signal in some automatic titrators.

All the foregoing methods of end-point evaluation are predicated on the assumption that the titration curve is symmetric about the equivalence point and that the inflection in the curve corresponds to this point. This assumption is perfectly valid, provided the participants in the titration react with one another in an equal molar ratio and also provided the electrode reaction is perfectly reversible. The former condition is lacking in many oxidation/reduction titrations; the titration of iron (II) with permanganate is an example. The curve for such titrations is ordinarily so steep; however, that failure to account for asymmetry results in a vanishingly small titration error. Silver nitrate is without question the most versatile reagent for precipitation titrations. A silver wire serves as the indicator electrode. For reagent and analyte concentrations of 0.1 M or greater, a calomel reference electrode can be located directly in the titration vessel without serious error from the slight leakage of chloride ions from the salt bridge.

This leakage can be a source of significant error in titrations that involve very dilute solutions or require high precision, however. The difficulty is eliminated by immersing the calomel electrode in a potassium nitrate solution that is connected to the analytic solution by a salt bridge containing potassium nitrate. Reference electrodes with bridges of this type can be purchased from laboratory supply houses. A theoretical curve to a potentiometric titration is readily derived. For example, the potential of a silver electrode in the argentometric titration of chloride can be described by;

$$E_{Ag} = E^0_{AgCl} - 0.0592 \log [Cl^-] = 0.222 - 0.0592 \log [Cl^-]$$

where E^0_{AgCl}, is the standard potential for the reduction of AgCl to Ag(s). Alternatively, the standard potential for the reduction of silver ion can be used:

$$E_{Ag} = E^0_{AgCl} - 0.0592 \log \frac{1}{[Ag^+]} = 0.799 - 0.0592 \log \frac{1}{[Ag^+]}$$

The former potential is more convenient for calculating the potential of the silver electrode when an excess of chloride exists, whereas the latter is preferable for solutions containing an excess of silver ion. Potentiometric measurements are particularly useful for titrations of mixtures of anions with standard silver nitrate.

Procedures

1- Weigh 0.1-0.2g of complexes then solve in 10ml water.
2- Fill the burette 0.1M of $AgNO_3$ solution
3- Read the potential of the solution after adding 0.5g ml of $AgNO_3$ solution
4- Find the end point of the titration and calculate the numbers of chlorides that are not in the coordination sphere.
5- Again weigh 0.1-0.2g of complexes into the 100ml of beaker then dry the complexes by the helping of 7-8ml 1M NaOH solution. In this procedure the complexes are decomposed. Repeat the steps 1 to 4 again. At the end of the titration the total number of chlorides in the complexes will be determined.
6- The difference between the number of chlorides in the step 5 and 6 gives the number of chlorides that are found in the primary coordination sphere.

Questions

1. Draw the structure of $Cu(II)(acac)_2$ complex.
2. Give a reaction mechanism for the formation of the acatylacetone complex.
3. The magnetic moment of $[Co(NH_3)_6]Cl_3$ is zero while that of $[Co(NH_3)_6Cl_2$ is 1.73BM. Why?
4. Record the infrared spectrum of the complex, $[Co(NH_3)_6]Cl_3$ as a liquid paraffin mull or as a KBr disc.
5. Draw the structure of the complex, $[Co(NH_3)_4Cl_2]Cl$.
6. Draw the more accurate structure of copper sulphate pentahydrate, $CuSO_4.5H_2O$.
7. Discuss the structures of $[Cu(NH_3)_4(H_2O)]^{2+}$ and $[Cu(NH_3)_4]^{2+}$ complex ions.
8. Explain the different type's methods of ammonia estimation.
9. What is the use of washing the ppt of $[Cu(NH_3)_4]SO_4.H_2O$ with methanol rather than water.
10. Draw the structure of the $Ni(C_4H_7O_2N_2)_2$ complex and indicate if intermolecular hydrogen bonding is possible.
11. Predict the magnetic momentum of $[Ni(NH_3)_6]Cl_2$.
12. Draw the shapes of $Ni(CO)_4$, $[Ni(CN)_4]^{2-}$ and $[Ni(NH_3)_6]^{2+}$ complexes.
13. Draw the geometrical structure of $[Fe(C_2O_4)_3]^{3-}$.
14. The magnetic moment of $K_4[Fe(CN)_6]$ is zero but that of $K_3[Fe(CN)_6]$ = 1.73 B.M. and $K_3[Fe(F)_6]$ = 5.9 B.M. Explain.
15. Fill the blanks:
- The ligand attached to the starting copper material used in the preparation of copper (II) complex is ....
- Name of the oxidant used in the preparation of potassiumtrioxalatoferrate(III) complex is ......
- The name of the iron (III) complex is ...., its formula is ....and the color is ....
- The color of Mohr's salt is ....and its formula is ....
- The molecular weights of: Copper (II) sulphate pentahydrate is ...., Potassium trioxalatochromate(III)trihydrate is ...., nickel chroridehexahydrate is ....,

chloropentaamminecobalt(III) chloride is........... and bis(dimethylglyoximato) nickel(II) is ....

16. What is the main role of the following reagents?
- Phosphoric acid in the analysis of chromium (III) ion
- Barium chloride solution in the analysis of sulphate
- Zinc dust and sulfuric acid in the analysis of iron (III) ion
- $AgNO_3$ solution and potassium persulphate in the analysis of chromium (III) ion
- Potassium Iodide in the analysis of copper (II) ion
- Ammonium chloride and conc. ammonia in the preparation of cobalt(III) complexes
- Addition of charcoal in the preparation of hexaamminecobalt(III)chloride

17. Discuss the background of the analysis of iron(III) ion.
18. Correct the following statements:
- The color of hexaamminenickel(II)chloride complex is green
- The gravimetric factor of Ni(II) in $Ni(DMG)_2$ is 0.4112
- The H_2O_2 is used to oxidize iron(III) to iron (II)
- In the titration process of permanganate ion and iron (II), the iron (II) loses five electrons and becomes oxidizing agent
- In the analysis of iron(III) ion, the zinc dust used to convert iron(III) to iron(II).

19. In titration analysis of copper(II) ion from tetraamminecopper(II)sulphate monohydrate, the weight of the sample was 0.3g and after titration with 0.05N $KMnO_4$ solution, the titration value at the end point was 25ml. Calculate the purity of copper in the prepared complex.

20. Briefly, discuss the calculations of sulphate ion in tetraammine copper(II)sulphate complex.

Index

Summary of certain chemical analysis of organic chemistry section

structures Alcohols	solubility H_2O	Oxidation green/blue	Nametal $\uparrow H_2$	Iodoform yellow	Borax phph	specific AAM
CH_3OH	W(pH=7)	+	+	-	-	Me-salicylate
CH_3CH_2OH	W(pH=7)	+	+	yellow	-	Ethyl acetate
$CH_3CH_2CH_2OH$	W(pH=7)	+	+	-	-	?
$CH_3CHOHCH_3$	W(pH=7)	+	+	yellow	-	?
$C_5H_{11}OH$	ss(pH=7)	+	+	-	+	?
$OHCH_2CHOHCH_2OH$	W(pH=7)	+	+	-	+	H_2SO_4
$C_6CH_{11}OH$	insoluble	+	+	-	-	?
$C_6H_5CH_2OH$	insoluble	+	+	-	-	dry heat

Aldehydes and Ketones	solubility	2,4-DNPH	Oxidation	Tollen	Fehling	specific
HCHO	W(pH=7)	+	green clor	Ag mirror	brwn-red ppt	Salicylic
CH_3CHO	W(pH=7)	+	green clor	Ag mirror	brwn-red ppt	Iodoform
$CCl_3CH(OH)_2$	W(pH=7)	yellow	green clor	Ag mirror	brwn-red ppt	Sodalime
C_6H_5CHO	insoluble	+	green clor	Ag mirror	brwn-red ppt	Oxid/ppt
CH_3COCH_3	W(pH=7)	yellow	-	-	-	Iodoform
$C_6H_{10}O$	insoluble	+	-	-	-	-
$C_6H_5COCH_3$	insoluble	+	-	-	-	Iodoform
$CH_3COCH_2COCH_3$	insoluble	+	-	-	-	Iodoform

Caboxylic acids	solubility	$NaHCO_3$	NuFeCl3	fluoriscein	phthalein	specific
HCO_2H	W(pH<7)	+ $\uparrow CO_2$	red-boold	-	-	-ve KMnO4
CH_3CO_2H	W(pH<7)	+ $\uparrow CO_2$	red-boold	-	-	EtOH/esterf
HO_2C-CO_2H	W(pH<7)	+ $\uparrow CO_2$	-	-	-	$CaCl_2$/pH7
$HO_2C-CHOH-CHOH-CO_2H$	W(pH<7)	+ $\uparrow CO_2$	yellow	-	-	Tollen's
$HO_2CCH_2C(OH)(CO_2H)CH_2CO_2H$	W(pH<7)	+ $\uparrow CO_2$	yellow	-	-	Dengen's
$HO_2C-CH_2-CH_2-CO_2H$	W(pH<7)	+ $\uparrow CO_2$	buff	red/green	-	
$C_6H_5-CO_2H$	NaOH+Acid	+ $\uparrow CO_2$	buff	-	-	-ve KMnO4
$o-HO_2C-C_6H_4-CO_2H$	NaOH+Acid	slow $\uparrow CO_2$	buff	intes green	bright red	
$C_6H_5-CH=CH-CO_2H$	NaOH+Acid	slow $\uparrow CO_2$	buff	-	-	KMnO4
$o-HO-C_6H_4-CO_2H$	NaOH+Acid	slow $\uparrow CO_2$	violet	-	bright red	MeOH/esterf

acid salts($M^+=NH_4^+,Na^+,K^+$)	solubility	$NaHCO_3$	$FeCl_3$	fluoresein	phthalein	specific
$CH_3CO_2^- M^+$	W(pH$\cong$ 7)	No CO_2 gas	red-boold	-	-	EtOH/esterf
$M^+ {}^-O_2C-CO_2^- M^+$	W(pH$\cong$ 7)	No CO_2 gas	-	-	-	$CaCl_2$
$MO_2C-CHOH-CHOH-CO_2^- M^+$	W(pH$\cong$ 7)	No CO_2 gas	yellow	-	-	Tollen's

^-O_2C-$CH_2C(OH)(CO_2H)CH_2$-CO_2^-	W(pH≅ 7)	No CO$_2$ gas	yellow	-	-	Dengen's
$M^{+-}O_2C$-CH_2CH_2-CO_2^- M^+	W(pH≅ 7)	No CO$_2$ gas	buff	red/ g.fluor	-	
C_6H_5-CO_2^- M^+	W(pH≅ 7)	No CO$_2$ gas	buff	-	-	-ve KMnO$_4$
$M^{+-}O_2C$-C_6H_4-CO_2^- M^+	W(pH≅ 7)	No CO$_2$ gas	buff	intes green	bright red	
C_6H_5-CH=CH-CO_2^- M^+	W(pH≅ 7)	No CO$_2$ gas	buff	-	-	KMnO$_4$
o-HO-C_6H_4-CO_2^- M^+	W(pH≅ 7)	No CO$_2$ gas	violet	-	bright red	MeOH/esterf

Phenols	NaHCO$_3$	FeCl$_3$	Liebrmn	Phthalein	Br$_2$/H$_2$O	specific
C_6H_5OH	No CO$_2$ gas	violet	green/blue	bright red	white	CHCl$_3$pink
m-HOC$_6$H$_4$OH	No CO$_2$ gas	violet/blue	green/blue	ints. green	yell-white	CHCl$_3$red
o-HOC$_6$H$_4$OH	No CO$_2$ gas	green-dark.	green/blue	?	no ppt	CHCl$_3$?
o or p-CH$_3$C$_6$H$_4$OH	No CO$_2$ gas	violet	?	?	?	CHCl$_3$red
α- or β-C$_{10}$H$_7$OH	No CO$_2$ gas	+	?	?	?	CHCl$_3$blue

Esters	solubility	NaHCO$_3$	NaOHcold	FeCl$_3$	Hydroxmat	specific
CH_3CO_2-C_2H_5	?	No ↑ CO$_2$	no NH$_3$	?	+	?
HOC$_6$H$_5$CO$_2$CH$_3$	-	No ↑ CO$_2$	no NH$_3$	?	+	?
CH$_3$COO-C$_6$H$_4$COOH	w(ph<7)	slow↑ CO$_2$	no NH$_3$	?	+	?

Amines	solub(HCl)	NaOH cold	FeCl$_3$	H$_2$O$_2$	Diazonium	specific
R-NH$_2$ (R$_2$N)	+	no NH$_3$	+	+	+	Azo dyes
C$_6$H$_5$NH$_2$	+	no NH$_3$	+	+	red dye	Azo dyes
p-NO$_2$C$_6$H$_4$NH$_2$	+	no NH$_3$	+	+	br. red dye	?
p-CH$_3$C$_6$H$_4$NH$_2$	+	no NH$_3$	green-blue	+	+	?
α or β-C$_{10}$H$_8$NH$_2$	+	no NH$_3$	+	+	+	?

Amine salts	solubility	NaHCO$_3$	FeCl$_3$	Diazonium	Ag$^+$/Ba^{+2}	specific
C$_6$H$_5$NH$_3$. HCl	W (pH< 7)	+	blue color	?	white ppt	AgNO$_3$
C$_6$H$_5$NH$_3$. H$_2$SO$_4$	W (pH< 7)	+	?	?	white ppt	BaCl$_2$

Amides	Soda lime	NaOH(Δ)	NaFusion	oxalic a.	FeCl$_3$	specific
NH$_2$CONH$_2$	NH$_3$ odor	NH$_3$ odor	+	white ppt	no color	Biuret
CH$_3$CONH$_2$	NH$_3$ odor	NH$_3$ odor	+	?	no color	Dry heat
C$_6$H$_5$CONH$_2$	NH$_3$ odor	NH$_3$ odor	+	?	no color	Dry heat

Chemicals needed for analysis in organic chemistry section

0	Classes	Compounds	structures	Requirement	Remarks
1	Alcohols	Methanol	CH_3OH	Experimental	
2	Alcohols	Ethanol	CH_3CH_2OH	theory	
3	Alcohols	1-Propanol	$CH_3CH_2CH_2OH$	Experimental	
4	Alcohols	2-Propanol	$CH_3CHOHCH_3$	Experimental	
5	Alcohols	Amyl alcohol	$C_5H_{11}OH$	Experimental	
6	Alcohols	Glycerol	$OHCH_2CHOHCH_2OH$	Experimental	
7	Alcohols	Cyclohexanol	$C_6CH_{11}OH$	theory	
8	Aldehydes	Formalin	$HCHO$	theory	
9	Aldehydes	Acetaldehyde	CH_3CHO	theory	
10	Aldehydes	Chloral hydrate	$CCl_3CH(OH)_2$	Experimental	
11	Aldehydes	Benzaldehyde	C_6H_5CHO	theory	
12	Ketones	Acetone	CH_3COCH_3	Experimental	
13	Ketones	Cyclohexanone	$C_6H_{10}O$	theory	
14	Ketones	Acetophenone	$C_6H_5COCH_3$	theory	
15	Ketones	Acetylacetone	$CH_3COCH_2COCH_3$	theory	
16	Caboxylic acid	Formic	HCO_2H	Experimental	
17	Caboxylic acid	Acetic	CH_3CO_2H	Experimental	
18	Caboxylic acid	Oxalic	$HO_2C\text{-}CO_2H$	Experimental	
19	Caboxylic acid	Tartaric	$HO_2C\text{-}CHOH\text{-}CHOH\text{-}CO_2H$	Experimental	
20	Caboxylic acid	Citric	$HO_2CCH_2C(OH)(CO_2H)CH_2CO_2H$	Experimental	
21	Caboxylic acid	Succinic	$HO_2C\text{-}CH_2\text{-}CH_2\text{-}CO_2H$	theory	
22	Caboxylic acid	Benzoic	$C_6H_5\text{-}CO_2H$	Experimental	
23	Caboxylic acid	Phthalic	$o\text{-}HO_2C\text{-}C_6H_4\text{-}CO_2H$	Experimental	
24	Caboxylic acid	Cinnamic	$C_6H_5\text{-}CH{=}CH\text{-}CO_2H$	theory	
25	Caboxylic acid	Salicylic	$o\text{-}HO\text{-}C_6H_4\text{-}CO_2H$	Experimental	
26	Na^+ or K^+ salts	M^+ Formate	$HCO_2^-\ M^+$	Experimental	
27	Na^+ or K^+ salts	M^+ Acetate	$CH_3CO_2^-\ M^+$	Experimental	
28	Na^+ or K^+ salts	M^+ Oxalate	$M^+\ {}^-O_2C\text{-}CO_2^-\ M^+$	theory	
29	Na^+ or K^+ salts	M^+ Tartarate	$MO_2C\text{-}CHOH\text{-}CHOH\text{-}CO_2^-\ M^+$	Experimental	
30	Na^+ or K^+ salts	M^+ Citrate	$^-O_2C\text{-}CH_2C(OH)(CO_2H)CH_2\text{-}CO_2^-$	Experimental	
31	Na^+ or K^+ salts	M^+ Succinate	$M^{+\ -}O_2C\text{-}CH_2CH_2\text{-}CO_2^-\ M^+$	Experimental	
32	Na^+ or K^+ salts	M^+ Benzoate	$C_6H_5\text{-}CO_2^-\ M^+$	Experimental	
33	Na^+ or K^+ salts	M^+ Phthalate	$M^{+\ -}O_2C\text{-}C_6H_4\text{-}CO_2^-\ M^+$	Experimental	
34	Na^+ or K^+ salts	M^+ Cinnamate	$C_6H_5\text{-}CH{=}CH\text{-}CO_2^-\ M^+$	theory	
35	Na^+ or K^+ salts	M^+ Salicylate	$o\text{-}HO\text{-}C_6H_4\text{-}CO_2\ M^+$	Experimental	
36	NH_4^+ salts	M^+ Acetate	$CH_3CO_2^-\ M^+$	theory	
37	NH_4^+ salts	M^+ Oxalate	$M^+\ {}^-O_2C\text{-}CO_2^-\ M^+$	Experimental	
38	NH_4^+ salts	M^+ Tartarate	$MO_2C\text{-}CHOH\text{-}CHOH\text{-}CO_2^-\ M^+$	theory	

190 General Physical Chemistry

#				
39	NH_4^+ salts	M^+ Citrate	$^-O_2C\text{-}CH_2C(OH)(CO_2H)CH_2\text{-}CO_2^-$	Experimental
40	NH_4^+ salts	M^+ Succinate	$M^{+-}O_2C\text{-}CH_2CH_2\text{-}CO_2^- M^+$	theory
41	NH_4^+ salts	M^+ Benzoate	$C_6H_5\text{-}CO_2^- M^+$	Experimental
42	NH_4^+ salts	M^+ Phthalate	$M^{+-}O_2C\text{-}C_6H_4\text{-}CO_2^- M^+$	theory
43	NH_4^+ salts	M^+ Cinnamate	$C_6H_5\text{-}CH=CH\text{-}CO_2^- M^+$	theory
44	NH_4^+ salts	M^+ Salicylate	$o\text{-}HO\text{-}C_6H_4\text{-}CO_2^- M^+$	theory
45	Phenols	Phenol	C_6H_5OH	Experimental
46	Phenols	. Resorcinol	$m\text{-}HOC_6H_4OH$	Experimental
47	Phenols	Catechol	$o\text{-}HOC_6H_4OH$	theory
48	Phenols	Cresols	o or $p\text{-}CH_3C_6H_4OH$	theory
49	Phenols	Naphthol	$\alpha\text{-}$ or $\beta\text{-}C_{10}H_7OH$	theory
50	Esters	Ethyl acetate	$CH_3COC_2H_5$	Experimental
51	Esters	Methyl Salicylate	$OHC_6H_4COCH_3$	Experimental
52	Esters	Aspirin	$CH_3COOC_6H_4CO_2H$	Experimental
53	Amines	Alkylamines	$R\text{-}NH_2$ (R_2N)	theory
54	Amines	Aniline	$C_6H_5NH_2$	Experimental
55	Amines	p- nitroaniline	$p\text{-}NO_2C_6H_4NH_2$	theory
56	Amines	Toluidine	$p\text{-}CH_3C_6H_4NH_2$	theory
57	Amines	Naphthylamine	α or $\beta\text{-}C_{10}H_8NH_2$	theory
58	Amine salts	Aniline . HCl	$C_6H_5NH_2. HCl$	Experimental
59	Amine salts	Aniline . H_2SO_4	$C_6H_5NH_2. H_2SO_4$	theory
60	Amides	Urea	NH_2CONH_2	Experimental
61	Amides	Acetamide	CH_3CONH_2	Experimental
62	Amides	Benzamide	$C_6H_5CONH_2$	theory

I) PHYSICAL properties

1) General properties

Condition	
Color	
Odor	

2) Solubility --

II) GENERAL REACTIONS AND TESTS

1) Action of dry heat

Inflammability	
Change of appearance	
Change of colour	
Change of odour	
Residue (After heating and addition of H_2SO_4)	

2) Action of soda lime

3) Action of 30% NaOH solution

4) Action of sodium carbonate solution

5) Reaction and coloration with Iron(III)chloride (Neutral medium)

6) Action of concentrated sulphuric acid

7) Detection of elements (N , S, X)

N	X			S
	Cl	Br	I	

8) FUNCTIONAL group reaction and identification

Test	Reagents	Observations	Indications (+ve or -ve)	Conclusion **AAM**
1				may be ...
2				
3				
4				

Special tests				

Major chemical equations:

III) FINAL Conclusion

The class is =

The compound is   =

Guide for the experiments in organic chemistry

1- Boiling and melting points determination:

	compound	mp	bp		compound	mp	bp
1-	Acetaldehyde	-	21	32-	Ethylmethyl ketone	-	80
2-	acetamide	81	-	33-	Glucose	83	-
3-	Acetanilide	115	-	34-	Heptane	-	89.4
4-	Acetic acid	16.6	118	35-	Hexane	-	69
5-	Acetic anhydride	-	139	36-	Hexanedioic acid	152	-
6-	Acetone	-	56.5	37-	Methanol	-	65
7-	Adipic acid	152		38-	Methoxyphenol -2	32	204
8-	Aniline	-	184	39-	Methylaniline -N	-	196
9-	Anthranilic acid	146		40-	Methylformate	-	31.5
10-	Benzaldehyde	-	179	41-	Methylphenol -3	-	191
11-	Benzamide	129	-	42-	Methylphenol -4	-	202
12-	Benzanilide	161	-	43-	Nephthaline	80.2	-
13-	Benzoic acid	122	250	44-	Nitrophenol -2	45	216
14-	Benzoin	137	-	45-	Nitrosalicylic acid -3	144	-
15-	Benzophenone	48	305	46-	Oxalic acid	101	-
16-	Bromobenzoic acid -m	156	-	47-	Pentanone -3	-	101
17-	Butanol -1	-	117.5	48-	Petrolium ether	-	40-60
18-	Butylamine	-	78	49-	Phenacetin	135	-
19-	Chlorobenzoic acid -o	140	-	50-	Phenol	-	181
20-	Cinnamic acid	133	300	51-	Phenyl acetate	-	196
21-	Cyclohexane	-	80.7	52-	Phenylacetone	27	216
22-	Cyclohexanol	-	161	53-	Phthalamide	221	-
23-	Cyclohexanone	-	156	54-	Phthalic anhydride	131	-
24-	Cylcohexene	-	83	55-	Salicylic acid	158	-
25-	Diethylaniline	-	215.5	56-	Styrene	-	156
26-	Dimethyl aniline N,N	-	193	57-	Succinic acid	188	-
27-	Ethoxyethyl acetate 2	-	156	58-	Thiourea	178	-
28-	Ethyl acetate	-	77	59-	Toluene	-	110.6
29-	Ethyl alcohol	-	78	60-	Trans-Cinnamic acid	133	-
30-	Ethyl benzoate	-	212	61-	Trichloromethane	-	76.7
31-	Ethylenediamine	-	117	62-	Urea	132.7	-
32-	Ethylmethyl ketone	-	80				

2- **Sublimation:** Camphor mp=179°, Naphthalene mp= 80°, Benzoic A. mp=122°

3- **Distillation:** Sodium Chloride, Ethyl alcohol, Cyclohexane

4- **Decolourzation:** Decolourizing Charcoal, Salicylic Acid, Naphthalene, Acetanilide

Experiment # 1: Cyclohexanol, H_3PO_4, H_2SO_4, $NaCl.Na_2CO_3$, Na_2SO_4

Experiment #1a: Conc. sulphuric acid, t-Amyl alcohol, NaOH.

Experiment #2: Cyclohexanol, Conc.HNO3, Potassium Permanganate, Potassium Dichromate, Ice sodium dichromate dihydrate, Acetic acid, K_2CO_3, NaCl, NaOH and ether.

Experiment #3: Conc. H_2SO_4, Sodium Dichromate ($Na_2Cr_2O_7.2H_2O$), cyclohexene

Experiment #3a: Adipic acid, KF, sodium chloride, ether, anhydrous sodium sulphate and $NaHCO_3$.

Experiment #4: Conc. H_2SO_4, N-butyl alcohol, sodium bromide dihydrate ($NaBr.2H_2O$), NaOH and anhydrous $CaCl_2$.

Experiment # 4a: Red Phosphorus, Ethanol (Preferably absolute), Iodine, 3% NaOH, anhydrous $CaCl_2$

Experiment #5: salicylic acid crystals, acetic acid anhydride, conc. sulfuric acid, phenol, ferric chloride, $NaHCO_3$, conc.HCl, acetate, petroleum ether

Experiment #6: n-amyl alcohol, glacial acetic acid, conc. H_2SO_4, anhydrous $MgSO_4$

Experiment # 7: sodium hydroxide, 95% ethanol, sodium chloride

Experiment #8: KOH, 95 % ethanol, acetophenone, benzaldehyde

Experiment #8a: benzaldehyde, sodium hydroxide, ethanol

Experiment #9: nitrobenzene, Zn (or $SnCl_2$) conc. HCl, NaOH, KOH pellets, ether.

Experiment #9a: aniline, acetic anhydride, conc. H_2SO_4, conc. HNO_3, NH_4OH, $.C_2H_5OH$, ice, CH_2Cl_2, activated charcoal, TLC.

Experiment #10: p-toluidine, charcoal, sodium acetate trihydrate, acetic anhydride, glacial acetic, ice, $MgSO_4$ hydrate, NH_4OH, ethanol, conc. H_2SO_4, Na_2CO_3, conc. HCl, $KMnO_4$, ether

Experiment # 11: $CuSO_4.5H_2O$, NaCl, sodium bisulphite, NaOH, p-toluidine, ice, conc. HCl, $NaNO_2$, starch-iodide paper, anhydrous sodium sulphate

Experiment #12: benzoic acid, absolute CH_3OH, small chips of porous porcelain, carbon tetrachloride, ice, $NaHCO_3$, $MgSO_4$, conc.H_2SO_4, conc. HNO_3, conc. HCl, NH_3, glacial acetic acid

Experiment #13: Ethyl acetoacetate, ethyl alcohol, phenylhydrazine

Experiment #14: Sodium, ethanol, diethyl malonate, conc. HCl.

Experiment #15: glycine, ethylene glycol, ice.

Experiment #16: pure aniline, conc. H_2SO_4, ether, conc. HCl, $ZnCl_2$, NaOH, anhydrous $CuSO_4$

Experiment #17: p-toluidine, acetylacetone, xylene, petroleum ether, conc. H_2SO_4, sodium carbonate

Experiment #18: Aniline, conc. HCl, Na_2SO_4 decahydrate, hydroxylamine.HCl, H_3PO_4

Experiment #19: Aniline, ethyl acetoacetate, glacial acetic acid, C_6H_6, diphenylether, petrol ether ($60-80°C$)

Experiment #20: cyclohexanone, glacial acetic acid, phenylhydrazine or phenylhydrazine.HCl, conc. H_2SO_4 anhydrous sodium acetate, conc. sulphuric acid

Experiment #21: benzoin, conc. HNO_3, petrolum ether, copper(II) acetate, NH_4NO_3, benzoin, CCl_4, acetic acid, Ethyl alcohol, benzil, Hexamine, glacial acetic acid, NH_4OH, pyridine

Experiment # 22: benzil, urea, NaOH, Ethyl alcohol, conc. HCl, methanol, spirit

References

1. R.C. Young, "Inorganic Syntheses", 2, 25(1946).
2. B.E. Bryant, W.C. Fernelius, "Inorganic Syntheses", 5, 188(1957).
3. A.I. Vogel, "A text book of Macro and Semi Micro Qualitative Inorganic Analysis", 4th edn., Longman, Green and CO LTD, London (1969).
4. Marr and Rockett, "Practical Inorganic Chemistry", Van Nostrand, (1972).
5. G. Marr and B.W. Rockett," Practical Inorganic Chemistry", 1st edn. Van Nostrand Reinold Company, London (1972).
6. A.I. Vogel, "Qualitative Inorganic Analysis", 6th edn., Longman Scientific, New York (1987) (Revised by G. Svehla).
7. F.G. Mann and B. C. Saunders, Practical Organic Chemistry, 4th Edn. (1990).
8. A. I. Vogel's Textbook of Practical Organic Chemistry, 5th Edn. (1996).
9. Robert de Levie "Principles of Quantitative Chemical Analysis, International ed. McGraw-Hill International (1997).
10. D. C. Harris, "Quantitative Chemical analysis", 6th edn., H.W. Freeman Company, New York (1999).
11. A. I. Vogel's Textbook of Quantitative Chemical Analysis, 5th Edn. (2002).
12. Gary D. Christian, Analytical Chemistry, 2nd Edn., Wiley International Edn. (1967).
13. The Merk Index , 9th Edn. (1976).
14. Pavia, Lampman and Kriz. Introdction to Organic chemistry techniques, W. B. Sounders Company (1976).
15. Shriner et.al., The systematic Identification of Organic Compounds, J.Wiley and Sons (1980).
16. F. Williamson, Organic Experiments, D.C. Health and Company (1983).
17. Vogel's, Practical Organic Chemistry, Longman Scientific and Technical (1991).
18. D.T. Plummer, An Introduction of Practical Biochemistry, Mc Graw Hill (1971).
19. Varley, Gowenlock and Bell, Practical Clinical Biochemistry, William Heinemann Medical Books Limited (1980).
20. N.W. Tietz, Fundamentals of Biochemistry, Saunsers (1982).
21. A. J. Pesce and L. A. Kaplan, Methods in Clinical Chemistry, Mosby (1987).

www.ingramcontent.com/pod-product-compliance
Lightning Source LLC
Chambersburg PA
CDIIW061020220326

41597CB00016BB/1749